TECHNOLOGY, PESSIMISM, AND POSTMODERNISM

TECHNOLOGY, PESSIMISM, AND POSTMODERNISM

Edited by

YARON EZRAHI
The Hebrew University of Jerusalem

EVERETT MENDELSOHN
Harvard University, Cambridge

and

HOWARD SEGAL
University of Maine, Orono

University of Massachusetts Press   Amherst

First published in 1994 in the series
Sociology of Sciences: A Yearbook, vol. XVII
Copyright © KLUWER ACADEMIC PUBLISHERS B. V. 1994
All rights reserved
Printed in the United States of America
LC 95–8315
ISBN 0–87023–977–5
First published in the United States of America in 1995
by the University of Massachusetts Press

Library of Congress Cataloging-in-Publication Data

Technology, pessimism, and postmodernism / edited by Yaron Ezrahi,
    Everett Mendelsohn, and Howard Segal.
        p.   cm.
    Originally published: Dordrecht; Boston: Kluwer Academic
    Publishers, © 1994, in series: Sociology of the sciences; v. 17.
    Includes bibliographical references.
    ISBN 0–87023–977–5 (alk. paper)
    1. Technology—Social aspects. 2. Technology and state.
    I. Ezrahi, Yaron. II. Mendelsohn, Everett. III. Segal, Howard.
    T14.5.T4463   1995
    306.4'5—dc20   95–8315
                                                            CIP

British Library Cataloguing in Publication data are available.

This book is published with the support and cooperation of the University
of Massachusetts Boston.

TABLE OF CONTENTS

ACKNOWLEDGMENTS

The idea for this volume gestated in a series of informal meetings held by the editors during the summer of 1989 when Professors Ezrahi and Segal were on the faculty of the Summer School at Harvard University. The Jerusalem/Tel Aviv site proposed by Ezrahi became possible when three Israel institutions agreed to become joint sponsors and the director of two of the institutions, Yehuda Elkana, joined in preparing for a conference of participants. The Hebrew University of Jerusalem's Edelstein Center for the History of Science, Tel Aviv University's Department of the History and Philosophy of Science, and the Van Leer Jerusalem Institute were generous financially and as hosts during the group's meetings. Originally scheduled to convene in April 1991, the conference was postponed until January 1992 due to the uncertainties caused by the lead into the Gulf War. The editors are deeply grateful to the three institutions in persevering with preparations and then graciously receiving the academic guests.

Bronwyn Mellquist who copy edited the papers and greatly improved their quality deserves appreciation from the editors and authors.

TECHNOLOGY, PESSIMISM, AND POSTMODERNISM:
INTRODUCTION

HOWARD P. SEGAL, FOR THE EDITORS

In November 1979 the Humanities Department of the University of Michigan's College of Engineering sponsored a symposium on "Technology and Pessimism." The symposium included scholars from a variety of fields and carefully balanced critics and defenders of modern technology, broadly defined. Although by this point it was hardly revolutionary to suggest that technology was no longer automatically equated with optimism and in turn with unceasing social advance, the idea of linking technology so explicitly with pessimism was bound to attract attention. Among others, John Noble Wilford, a *New York Times* science and technology correspondent, not only covered the symposium but also wrote about it at length in the *Times* the following week.

As Wilford observed, "Whatever their disagreements, the participants agreed that a mood of pessimism is overtaking and may have already displaced the old optimistic view of history as a steady and cumulative expansion of human power, the idea of inevitable progress born in the Scientific and Industrial Revolutions and dominant in the 19th century and for at least the first half of this century." Such pessimism, he continued, "is fed by growing doubts about society's ability to rein in the seemingly runaway forces of technology, though the participants conceded that in many instances technology was more the symbol than the substance of the problem."[1]

To be sure, technology had its defenders at the 1979 symposium, particularly Melvin Kranzberg, one of the founders of the academic discipline of the history of technology, and Samuel Florman, the popular writer and practicing civil engineer. Kranzberg argued that technology's very success in the most industrialized societies ironically provided them with the "luxury" of being able to criticize technology's alleged excesses and to call for such non-technological objectives as social justice and environmental preservation. Until recently, he claimed, sheer survival needs and persistent scarcity precluded those considerations. Florman, for his part, invoked his own familiar "tragic view" of life and asked that we recognize inherent and unchanging human frailties and abandon unrealistic expectations for what technology might accomplish – thereby simultaneously and cleverly evading the issue of what difference it makes for admittedly flawed humans to live amid varying degrees of technological development over time.[2] Technology's foremost defenders, however, were not present: the University's engineering faculty, who not only boycotted the symposium but a year later succeeded in abolishing the Humanities Department's annual lecture series on "Technology and Values," the venue for the symposium, and a few years later succeeded in closing the Humanities Department itself. For the engineers, any serious criticism of technology was heresy.

That 1979 symposium, which I helped to organize, was the personal inspiration for the 1992 Israel workshop that led to this volume in *The Sociology of the Sciences Yearbook* series. The general topic of technological pessimism seemed worthy of reconsideration in the 1990s and the annual international workshop a most appropriate forum for such reconsideration. It is a measure of the degree of change in common perceptions of technology since 1979 that, by 1990, when the solicitation of papers for the Israel workshop began, technological pessimism could be taken for granted. The workshop organizers did not need to worry about including – or offending – true believers in technological optimism. Nor were the various institutions cosponsoring the workshop outraged by the very topic in the manner of the Michigan engineering faculty.

Ironically, what most threatened the workshop was itself a source of one form of technological success: the Persian Gulf War, whose outbreak forced postponement of the workshop from April 1991 till January 1992. The association of the Gulf War with renewed technological prowess on the part of the United States is explored in my own contribution.

The Gulf War, however, was only the immediate backdrop to the overarching concerns outlined in the original call for papers. "A faith in the powers of knowledge and technology to ameliorate human life and solve the basic problems of modern society has been one of the central features of Western culture. This faith nourished Western utopias and ideologies for at least three centuries." Until recently, the workshop solicitation continued, "the massive application of rational knowledge and technical skills" was generally deemed the panacea to bring about qualitatively superior societies, not just in the West but throughout the world. (The workshop solicitation conceived of technology broadly as consisting of forms and knowledge and technical skills as well as of machines and tools themselves; by now few students of technology would question this conception.) Earlier in the twentieth century, this faith was challenged by Fascism's "reactionary modernism" – to use the phrase of contributor Jeffrey Herf – which employed technology in the "politics of unreason." After World War II this faith manifested itself more "positively" in the innumerable efforts in the West and elsewhere to deploy technology in virtually every sphere of public policy, from defense to education to energy to health to communications and transportation. Ideology and politics itself were predicted to decline as supposedly neutral information, rational analysis, and technical skills took their places. Enlightened liberal democracies would thereby flourish while "underdeveloped" countries would eventually follow the same route. Totalitarian communist regimes would, of course, pose the fiercest competition for the loyalties of the latter nations, but they, too, would someday see the light (or be forced into submission). After all, even Lenin had tried to adopt Henry Ford's assembly line techniques and Frederick Taylor's scientific management schemes back in the 1920s. Conventional wisdom had it that the supremacy of the powers of the rational knowledge and technical skills to improve human life rested on their alleged universality and efficiency alike.

Somehow, and despite any number of spectacular technological – and scientific – breakthroughs, there has emerged in recent decades a declining trust in the social and political applications of knowledge and, more broadly, in technical expertise itself. (The workshop implicitly recognized the differences between science and technology but did not exclude science from its consideration where appropriate, as several of the contributors make clear; but technology remained the focal point.) Indeed, the triumphs of science and technology that once generated almost universal praise increasingly generate distrust or at best ambivalence on the part of the public. The prospect of autonomous technology, of technology without human control, has become a paramount public concern regardless of any statistical and other quantitative evidence to the contrary. In turn public reaction to the very notion of "fail-safe" technological systems has shifted from reassurance to suspicion and even mockery. Today, those public officials such as President Jimmy Carter who use their technical training (as in his case, in nuclear engineering) to claim apolitical administrative expertise – much as President Herbert Hoover, a mining engineer by profession, claimed during his term – are widely belittled as supreme political opportunists, which was not the case during the Hoover Presidency. Appeals for support for public policies, much less for votes, on the basis of engineering and related expertise thereby fall on ever more deaf ears. If anything, as Yaron Ezrahi's *The Descent of Icarus* (1990) elaborates, the hitherto common reliance on science and technology to justify the modern state's interventions in most areas of public (and private) life is increasingly in jeopardy.

No less important, technological pessimism has become an integral part of the emerging culture of postmodernism. Within that cultural hierarchy, technology itself may be assuming a declining status amid a growing disenchantment with material success and with all forms of social and political engineering. In the postmodern critique, the project of modernity itself is under sharp attack. Nowhere is this disenchantment more evident than with regard to contemporary environmental concerns. The increasingly shared sense, as Rosalind Williams put it in a conference paper to be published in *Science in Context*, that "we live in a ruined world" of our own making, is a fundamental source of technological pessimism. Where earlier visionaries, religious and secular alike, equated environmental reform with activity, we today equate it with restraint. Not surprisingly, postmodern redefinitions of self, society, history, and politics do not grant technology the dominant role that it enjoyed under the Enlightenment vision. And that in turn is forcing a redefinition of technology itself as a mode of human action and new notions of what "fulfillment" with and without technology means.

The prospect of a "ruined world" inevitably raises questions about the possible irreversibility of technological "progress" not just in the environment but everywhere. That such questions should arise in the first place surely bespeaks the postmodern sensibility, as does the growing belief that the irreversibility of what had frequently been deemed not merely technological but outright social advances is a hallmark of our age. There is, of course, still the "technological fix," as espoused most notably by nuclear engineer Alvin Weinberg. As Weinberg argued in 1966, when skepticism about unceasing technological progress was beginning to take hold in the United States and other highly industrialized countries, technology could readily find shortcuts – or technological quick fixes – to the solution of social problems. These would supposedly avoid the likely futile efforts to try changing certain people's undesirable values and mores to socially and politically acceptable ones. His examples included utilizing the technologies of energy, mass production, and automation to eliminate poverty in affluent societies; IUD's to reduce overpopulation in "underdeveloped" countries; nuclear-powered desalination to provide fresh water in needy areas and serve as a basis for solving the Arab-Israeli conflict; cheap computers to supplement or replace teachers in elementary schools; free air conditioners and free electricity to reduce riots in urban ghettos; safe cars to reduce accidents, injuries, and deaths everywhere; and not least and most obvious, atomic weapons to reduce the possibility of war anywhere. That all of these were, contrary to Weinberg, not alternatives to, but rather themselves examples of, social engineering reflects the persistence of Enlightenment dreams. Indeed, despite his claim of being a tough-minded realist painfully aware (like Florman) of eternal human flaws and inadequacies, Weinberg is paradoxically as utopian as those whom he dismisses as romantics, but the source of his utopianism is the power of technology, not the power of persuasion.[3]

None of the participants in the workshop could be accused of harboring utopian fantasies of this or any other kind, but none was likewise as pessimistic about technology and the future as, say, Jacques Ellul, whose influential writings were invoked more than once. If everyday life actually resembled the world he describes in *The Technological Society* (1954), one in which there is literally "no exit" from modern technology's overwhelming and unceasing grasp, then the logical alternatives for sensible souls would be resignation or suicide – and no more time spent on reading such depressing works as his own. Yet even Ellul, one suspects, retains at least slight hope that, precisely by reading such works as his, positive change of some kind will somehow and somewhere occur. Like the most naive technological utopian, this starkest of technological pessimists surely aspires to a mass audience if not a devoted following. And as Yaron Ezrahi observes in his contribution, technological pessimism is that much harder for the poor and the powerless to accept than for the rest of any society. The former have so little to live for and so few rewards for their daily travails that the very notion of a world out of virtually everyone's control and in the hands of an autonomous technological monster, the world Ellul describes as contemporary reality, is impossible for them to acknowledge. Better for them the utopian – and paternalistic – fantasies of Weinberg.

A far healthier prescription, however, is that outlined in Leo Marx's contribution. It is notable in suggesting ways of not merely recognizing but also coping with technological pessimism. As he argues, the notion of modernization so long identified with increasing rational knowledge and technical skills at the expense of virtually any environmental concerns is gradually being redefined, reconstituted, and even completed to include notions of nature and of wholeness that are not archaic, not outmoded, contrary to the conventional wisdom of so much of this and earlier centuries. Marx seeks a new or revived pastoralism that accepts the centrality of the society-nature connection in a nonexploitative relationship; that prefers equitable sufficiency to maximization of resources, goods, and services; that rejects crass economic criteria (such as the Gross National Product) as the principal basis for public policy; that accepts lower standards of living as strictly quantitative measures in return for a higher quality of life and a greater happiness. Marx offers, in effect, a kind of "appropriate technology" for the most industrialized countries and a different vision of modernity for the entire world.

One need not accept Marx's position to recognize that, as he and other contributors suggest, technological pessimism can and often does represent a progressive shift in worldview. The end of the Enlightenment dream may thus be a cause of celebration as much as of despair, at least for some. Not only in the environmental arena, but in others as well, such as work and leisure, hidden yearnings for wholeness and for meaning have, in the eyes of many – and not just intellectuals – for too long now been repressed or outright dismissed as backward and romantic by uncritical advocates of science and technology. Can and should we continue to ignore these yearnings? And what positive equivalent for nostalgia, with its negative connotation, can we substitute? Can we speak and write without embarrassment of the "reenchantment" of the world?

Everett Mendelsohn in his contribution focuses on the rebellions of the 1960s and 1970s and the call by such diverse critics as Lewis Mumford, Herbert Marcuse, and Theodore Roszak to reunite fact and value in the sciences and technology. Having identified "instrumental knowledge" or "objective knowledge" as the source of a science and technology increasingly divorced from human life and human need, they and others called for a re-invention of the human relationship to nature. Ido Yavetz's contribution cautions us not to overgeneralize about the extent of technological optimism in the past. By looking at a late-nineteenth-century British controversy over lightning rod protection, he reveals not "simple arguments for or against technology" but rather a "complex range, in which different, often competing notions of progress appear, with some people advocating certain technologies as opposed to others." If Yavetz's case study is at all representative, then it follows that progress meant different things to different people in Victorian England as presumably elsewhere and that, as other scholars have long argued, much depends on precisely which technology is at issue. Whether, however, as Yavetz wonders, this earlier qualified optimism means that contemporary technological pessimism is merely the continuation of a "well-established tradition of assimilation by criticism of new technologies into our ever changing social and cultural context" is quite another question. To discover some similarities between a century ago and today is not necessarily to discover overriding continuities between then and now.

For many critics of contemporary technology, the kind of ambivalence which society seems often to use to accommodate technology represents the initial formulation of a new notion of progress, even a new worldview, and the qualification of the irreversibility of now questionable technological progress noted above. For a defender of contemporary technology, however, such as historian of technology Jean Gimpel, any rejections of unadulterated technological advance represent the decline of civilization and confirmation of the cyclical nature of history itself. In *The Medieval Machine* (1976) Gimpel boldly compares the rejection of certain technological improvements by late medieval Europe with the rejection of the supersonic transport by the United States Congress in 1971. Disillusioned by this failure to fund what certainly proved to be a financial if not necessarily a technological fiasco, he predicts a technological dark age for the United States akin to that of the Middle Ages (which, ironically, by his own account, was anything but dark technologically.) Other nations will have to be responsible for future technological leadership. For Gimpel, then, any deviation from the "technological imperative," the notion that whatever can be achieved technologically ought to be achieved, is simply unacceptable.

The alleged stakes for humankind's future represented by these different positions in turn raises the fundamental issue addressed by several contributors: what and whose political, economic, and ideological ends are being served by technological pessimism or optimism? Robert Pippin's analysis and critique of "The Notion of Technology as Ideology" illuminates the complexity of these concerns. As he argues, "the right metaphor for understanding the extraordinary and potentially distorting appeal of technological power in modernity is not a hunt for hidden origins, or a delineation of geographical boundaries, but attention to the *context*, the historical moment when mastery in general would have seemed, with some historical urgency, an unavoidable *desideratum*." In this broader context the appeal of technological mastery will clearly be separated from "some sort of Faustian bargain, prompted by hubris, narrow class interest, confusion about different domains of rationality, or as a lust for power."

Along with renewed environmental consciousness, renewed political consciousness is a hallmark of technological pessimism. As several contributions make clear, and as the workshop's own solicitation implied, politics has not been replaced by technology and never will be. As Ezrahi contends in his paper, technological improvements offer no lasting escape from politics, and twentieth-century claims that they do, as exemplified by liberal political scientists like Robert Lane as well as by liberal technocrats like Alvin Weinberg, are untenable. Because of its inherent contradictions and elusiveness, the public interest can never be defined technologically. In fact, the viability of the liberal democratic institutions that Lane and Weinberg endow with near-utopian qualities is itself a major source of technological pessimism for many others. Not surprisingly, the revival of politics amid technological pessimism has increasingly been at the local or "grass-roots" levels. It is no coincidence that a recent semi-popular study of local opposition to various technological developments imposed from above – the so-called "not in my back yard phenomenon" – is subtitled *Community Defiance and the End of Technological Optimism*.[4] Yet Ezrahi reminds us that the localization of technological issues can lead as often to fanaticism as to democracy. He and others reminded the workshop that, in the name of preserving the environment, so-called "eco-fascism" is a possible threat to the United States and other societies as individuals and groups put environmental concerns above human life and well-being.

As noted at the outset, the 1991 Gulf War presents an intriguing case study of technological pessimism and optimism alike: pessimism concerning the war itself and its many civilian and military casualties plus untold damage to the regional environment; optimism concerning the comparative brevity of the conflict and the (alleged) accuracy of its high tech weapons system. Yet a fundamental point made by several of the workshop participants is that technological problems themselves are rarely the real issue. This may well be the ultimate irony of the failure of Enlightenment dreams and the profoundest difference between traditional technological optimism and current and future technological pessimism. Herf states his position bluntly: "After Auschwitz, it was not technology but human beings who stood condemned." For him, technological pessimism in twentieth-century Germany has been and, presumably, in other societies may yet be "one of many varieties of determinism and the avoidance of human agency" that offers nothing helpful in trying to solve problems caused by the use and abuse of man-made technologies. If Nazis could blame technology rather than themselves for their crimes against humanity, so, logically, might latter-day totalitarians. Except briefly after 1945, this happily has been the case neither in postwar West Germany nor, for different reasons, in East Germany. Yet the vision of autonomous technology inspired by Ellul and others of very different political persuasions may still lend itself to the kind of evasion of political and moral responsibility described by Herf. And Herf locates new versions of technological pessimism in the (West) German Left as epitomized by the Greens and the peace movements of the 1980s in their respective predictions of environmental and nuclear apocalypses. (Herf's earlier work on the antidemocratic Right in Weimar Germany and then under the Nazis *Reactionary Modernism* [1984] also cautions us not to link cultural pessimism automatically with technological pessimism, for leading intellectuals in both regimes were as enthusiastic about technological advance as they were pessimistic about the West and modernity; perhaps technological pessimism could thus coexist elsewhere today with cultural optimism in some combination.)

Leo Marx has identified another, no less important, dimension of this reduction of the significance of purely technological problems amid technological pessimism. As he suggests in a recent review of a book of essays in honor of Melvin Kranzberg, the paradoxical result of ever greater knowledge and understanding of technology is "to cast doubt on the rationale for making 'technology,' with its unusually obscure boundaries, the focus of a discrete field of specialized historical (or other disciplinary) scholarship." Quoting Martin Heidegger, as did others in the workshop, Marx asked, "Is it not probably that 'the essence of technology is by no means anything technological'?"[5] And if that be the case, what is the deepest significance of technological pessimism?

Hence the futility of appeals by engineers and other technical experts to create "a new image," in the words of the chairman of a major aerospace corporation writing recently in *Newsweek*. "To a considerable segment of the public," Norman Augustine laments, "the word 'technology' conjures up images of Chernobyl, Bhopal, and the Challenger. Too often, technology is perceived as the problem rather than the solution; as something to be avoided rather than embraced." His own solution is the participation especially of engineers in public debates over technological issues and the end of engineers' characteristic "voiceless and invisible" pose when pertinent "matters of public policy are debated." The example he cites is the supersonic transport, the rejection of which by Congress still outrages him as much as it once did Gimpel. He suggests that if only engineers had made their naturally supportive position known, that alone would have justified a questionable enterprise. Augustine calls on engineers to recognize their "obligation in this regard if we are to fulfill our potential to be among the guardians of the world's highest quality of life" (wherever that may be; he doesn't specify).[6] The gap between this stance, however heroic and well-intentioned, and the nature and degree of technological pessimism outlined above and detailed in the contributions that follow, is so great as to be cosmic.

Yet Augustine is worth quoting just because such desperate pleas for respect – and power – reflect how far the world has come in recent decades from the self-confidence in technology and in progress exemplified by the writings of Lane and Weinberg. The erosion of faith in what Kenneth Keniston in his conference paper terms "the engineering algorithm" – the core belief that "the relevant world can be defined as a set of problems, each of which can be solved through the application of scientific theorems and mathematical principles"[7] – cannot be repaired by the participation in the political process by the most passionate and persuasive of engineers in the United States and elsewhere. If anything, their political protests would surely be countered by those already active in environmental, disarmament, antinuclear, and other movements who are no longer impressed by either technical expertise or its questionable claims to objectivity. Nor, as I argue in my paper, would the greatest success of the contemporary technological literacy crusade prevent further erosion, particularly if technology's mixed blessings are taught and appreciated. Rather, as Ezrahi observes, the changes in the political process produced by the demands of women, minorities, and other persons for greater involvement and representation may ultimately have been more significant for the growth of technological pessimism (or technological ambivalence) than any writings on technological pessimism – present company excluded, perhaps.

Notes

1. John Noble Wilford, "Scholars Confront the Decline of Technology's Image," *The New York Times*, November 6, 1979, C1.
2. See Melvin Kranzberg, "Technology: The Half-Full Cup," and Samuel G. Florman, "Technology and the Tragic View," in *Alternative Futures: The Journal of Utopian Studies*, 3 (Spring 1980), 5–18 and 19–30. These are revised versions of their respective symposium papers.
3. See Alvin M. Weinberg, "Can Technology Replace Social Engineering?" *University of Chicago Magazine*, 59 (October 1966), 6–10.
4. See Charles Piller, *The Fail-Safe Society* (New York: Basic Books, 1991).
5. Leo Marx, review of *In Context: History and the History of Technology – Essays in Honor of Melvin Kranzberg*, in *Technology and Culture*, 32 (April 1991), 394–396.
6. Norm Augustine, "My Turn: Engineering a New Image," *Newsweek*, February 10, 1992, 13.
7. Kenneth Keniston, "Trouble in the Temple: The Erosion of the Engineering Algorithm" (unpublished conference paper), 10.

THE IDEA OF "TECHNOLOGY" AND POSTMODERN PESSIMISM

LEO MARX

Massachusetts Institute of Technology

"The factor in the modern situation that is alien to the ancient regime
is the machine technology, with its many and wide ramifications."
— Thorstein Veblen (1904)[1]

"Technological Pessimism" and Contemporary History

"Technological pessimism" may be a novel term, but most of us seem to
understand what it means.[2] It surely refers to that sense of disappointment,
anxiety, even menace, that the idea of "technology" arouses in many people
these days. As the editors of this volume note, however, there also is something
paradoxical about the implication that technology is somehow responsible for
today's widespread social pessimism. The modern era, after all, has been marked
by a series of "spectacular scientific and technological breakthroughs"; we are
reminded of the astonishing technical innovations of the last century in, say,
medicine, chemistry, aviation, electronics, atomic energy, space exploration, or
genetic engineering. Isn't it odd, then, to attribute today's widespread gloom to
the presumed means of achieving all those advances: an abstract entity called
"technology"?

A predictable rejoinder, of course, is that in recent decades that same entity
also has been implicated in a spectacular series of disasters. One thinks, for
example, of Hiroshima, the nuclear arms race, the American war in Vietnam,
Chernobyl, Bhopal, the Exxon oil spill, acid rain, global warming, or ozone
depletion. Each of these events was closely tied to the use or the misuse, the
unforeseen consequences or the malfunctions, of relatively new and powerful
science-based technologies. Even if we fully credit the technical achievements
of modernity, their seemingly destructive social and ecological consequences
(or side effects) have been sufficiently conspicuous to account for much of
today's "technological pessimism."

If we are ambivalent about the effects of technology in general it is because,
for one thing, it is so difficult to be clear about the consequences of particular

From M. R. Smith and L. Marxs (eds.): *Does Technology Drive History?: The Dilemma of Technological Determinism* (MIT Press, 1994).

kinds of technical innovation. Take, for example, modern advances in medicine and social hygiene, perhaps the most generally admired realm of science-based technological improvement. Today it nonetheless is said that those alleged advances are as much a curse as a blessing. In privileged societies, to be sure, medical progress has resulted in curbing or eliminating many diseases, prolonging life, and lowering the death rate; in large parts of the underdeveloped world, however, those very achievements have set off a frightening, possibly catastrophic growth of population, with all its grim ramifications. Is it any wonder, considering the plausibility of that gloomy "Malthusian" view, that advances in medicine may issue in pessimism as well as optimism?

On reflection, however, such inconclusive assessments seem crude and ahistorical. They suffer from a "presentist" fallacy like that which casts doubt on the results of much public-opinion polling. It is illusory to suppose that we can isolate for analysis the immediate, direct responses to specific innovations. Invariably people's responses to the new – to changes effected by, say, a specific technical innovation – are mediated by older, preexisting attitudes. Whatever their apparent spontaneity, such responses usually prove to have been shaped by meanings, values, and beliefs that stem from the past. A group's responses to an instance of medical progress cannot be understood, for example, apart from the historical context, and more specifically, the expectations generated by the belief that modern technology is the driving force of progress.

Technological Pessimism and the Progressive World Picture

The current surge of "technological pessimism" in advanced societies is closely bound up with the central place accorded to the mechanic arts in the progressive world picture. That image of reality has dominated Western secular thought for some two and a half centuries. Its nucleus was formed around the late eighteenth-century idea that modern history itself is a record of progress. (In the cultures of modernity, conceptions of history serve a function like that served by myths of origin in traditional cultures: they provide the organizing frame, or binding meta-narrative, for the entire belief system.) Much of the extravagant hope generated by the Enlightenment project derived from trust in the virtually certain expansion of new knowledge of and enhanced power over nature. At bottom this historical optimism rested upon a new confidence in humankind's capacity, as exemplified above all by Newtonian physics and the new mechanized motive power, to discover and put to use the essential order – the basic "laws" – of Nature. The expected result was a steady, cumulative improvement in the conditions of life. What requires emphasis here, however, is that advances of science and the practical arts were singled out as the primary agents of progress.

In the discourse of the educated elite of the West between 1750 and 1850, the idea of progress often seems to have been exemplified by advances in scientific knowledge; at more popular levels of culture, however, progress more often was exemplified by innovations in the familiar practical arts. Although "science" was identified with a body of certain, mathematically verifiable knowledge – abstract, intangible, and recondite – the mechanic arts were associated with the commonsense practicality of everyday artisanal life as represented by tools, instruments, or machines. Nothing provided more tangible, vivid, compelling icons for representing the forward course of history than recent mechanical improvements like the steam engine.[3]

A recognition of the central part that the practical arts were expected to play in carrying out the progressive agenda is essential for an understanding of today's growing sense of technological determinism – and pessimism. The West's dominant belief system turned on the idea of technical innovation as a primary agent of progress. Nothing in the Enlightenment world picture prepared its adherents for the shocking series of twentieth-century disasters linked with, and often apparently caused by, the new technologies. On the contrary, with the increasingly frequent occurrence of such disasters since Hiroshima, more and more people in the "advanced" societies have had to consider the possibility that the progressive agenda, with its promise of limitless growth and a continuing improvement in the conditions of life for everyone, has not been, and perhaps never will be, realized.[4] The sudden dashing of those long-held hopes surely accounts for much of today's widespread technological pessimism.

All this may be obvious, but it does not provide an adequate historical explanation. To understand why today's social pessimism is so closely bound up with the idea of technology, it is necessary to recognize how both the character and the representation of technology changed in the nineteenth century. Of the two major changes in the character of technology, one was primarily material or artifactual: the introduction of mechanical (later, chemical and electrical) power and the consequent development of large-scale, complex, hierarchical, and centralized systems like the railroad or electric power systems. The second, related development, was ideological: the atrophy of the Enlightenment idea of progress directed toward a more just, republican society, and its gradual replacement by a politically neutral, technocratic idea of progress whose goal is the continuing improvement of technology. The improvement of technology came to be seen as the chief agent of change in an increasingly deterministic view of history.

Understanding these changes is complicated, however, by the fact that the most fitting language for describing them also came into being as a result of, and indeed largely in response to, these very changes.[5] The crucial case is that of "technology" itself. To be sure, we intuitively account for the currency of the word in its broad modern sense as an obvious reflex of the increasing proliferation in the nineteenth century of new and more powerful machinery. But this is not an adequate historical explanation. It reveals nothing about the preconditions, the conceptual or expressive needs, that called forth this new word. Such an inquiry is not trivial, nor is it "merely" semantic. The genesis of this concept, as embodied in its elusive prehistory, is a distinctive feature of the onset of modernity.[6] Not only will it illuminate the rise of "technological pessimism," it will help us to see that that phenomenon, far from being a result of recent events, had its origin in the very developments that called into being, among other salient features of modernity, the idea of "technology."

The Changing Character of the "Mechanic Arts" and the Invention of "Technology"

When the Enlightenment project was emerging after 1750, the idea of "technology" in today's broad sense of the word did not yet exist. For another century or so, the artifacts, the knowledge, and the practices later to be embraced by "technology" would continue to be thought of as belonging to a special branch of the arts known as the "mechanic" (or "practical" or "industrial" or "useful") – as distinct from the "fine" (or "high" or "creative" or "imaginative") – arts. Such terms were then the nearest available approximations of today's abstract noun "technology"; they referred to the knowledge and practice of the crafts. By comparison with "technology," the "practical arts" and its variants constituted a more limited and limiting category. If only because it was explicitly designated as one of several subordinate parts of something else, such a specialized branch of art was, as compared with the tacit distinctiveness and unity of "technology," inherently belittling. Ever since antiquity, moreover, the habit of separating the practical and the fine arts had served to ratify a set of overlapping invidious distinctions between things and ideas, the physical and the mental, the mundane and the ideal, female and male, making and thinking, the work of enslaved and of free men. This derogatory legacy was in some measure erased, or at least masked, by the more abstract, neutral word "technology." The term "mechanic arts" calls to mind men with soiled hands tinkering with machines at a workbench, whereas "technology" conjures up images of clean, well-educated technicians gazing at dials, instrument panels, or computer monitors.

These changes in the representation of technical practices were made in response to a marked acceleration in the rate of initiating new mechanical or other devices and new ways of organizing work. During the early phase of industrialization (c. 1780–1850 in England, c. 1820–1890 in the United States), the manufacturing realm had been represented in popular discourse by images of the latest mechanical inventions: water mill, cotton gin, power loom, spinning jenney, steam engine, steamboat, locomotive, railroad "train of cars", telegraph, or factory. The tangible, manifestly practical character of these artifacts matched the central role accorded to instrumental rationality and its equipment as chief agent of progress. Thus the locomotive (or "iron horse") often was invoked to symbolize the capacity of commonsensical, matter-of-fact, verifiable knowledge to harness the energies of nature. It was routinely depicted as a driving force of history. Or, put differently, these new artifacts represented the innovative means of arriving at a socially and politically defined goal. For ardent exponents of the rational Enlightenment, the chief goal was a more just, less hierarchical, republican society based on the consent of the governed.

As this industrial iconography suggests, the mechanic arts were widely viewed as a primary agent of social change. These icons often were invoked with metonymical import to represent an entire class of similar artifacts, such as mechanical inventions; or the replacement of wood by metal construction; or the displacement of human, animal, or other natural energy sources (water or wind) by engines run by mechanized motive power; or some specific, distinctive feature of the era ("the annihilation of space and time," "The Age of Steam"); or, most inclusively, its general uniqueness (the "Industrial Revolution"). Thus when Thomas Carlyle announced at the outset of his seminal 1829 essay, "Signs of the Times," that if asked to name the oncoming age, he would call it "The Age of Machinery," he was not merely referring to actual, physical machines, or even to the fact of their proliferation.[7] He had in mind a radically new kind of ensemble typified in, but by no means restricted to, actual mechanical artifacts. The machine, as invoked by Carlyle (and soon after by many others), had both material and ideal (mental) referents; it simultaneously referred to (1) the "mechanical philosophy," an empirical mentality associated with Descartes and Locke, and the new science, notably Newtonian physics, that it had generated; (2) the new practical or industrial arts (especially those using mechanized motive power); (3) the systematic division of labor (the workers as cogs in the productive machinery); and (4) a new kind of impersonal, hierarchical, or bureaucratic organization, all of which could be said to exhibit the power of "mechanism." Carlyle's essay is an early, eloquent testimonial to the existence of a semantic void, and the desire for a more inclusive, scientistic, and distinctive conception of these new human powers than was signified by the most inclusive term then available, "the mechanic arts."

During the nineteenth century, discrete artifacts or machines were replaced, as typical embodiments of the new power, by what later would come to be called "technological systems."[8] It is evident in retrospect that the steam-powered locomotive, probably the leading nineteenth-century image of progress, did not adequately represent the manifold character or complexity of the mechanic art of transporting persons and goods by steam-powered engines moving wagons over a far-flung network of iron rails. To represent such complexity, that image (of a locomotive) was no more adequate than the term "mechanic art." As Alfred Chandler and others have argued, the railroad probably was the first of the large-scale, complex, full-fledged technological systems.[9] In addition to the engines and other material equipment (rolling stock, stations, yards, signalling devices, fuel supplies, the network of tracks), a railroad comprised a corporate organization, a large capital investment, and a great many specially trained managers, engineers, telegraphers, conductors, and mechanics. Because a railroad operated over large geographical areas, twenty-four hours a day, every day of the year, in all kinds of weather, it become necessary to develop an impersonal, expert cohort of professional managers, and to replace the traditional organization of the family-owned and operated firm with that of the large-scale, centralized, hierarchical, bureaucratic corporation.

Between 1870 and 1920 such large complex systems became a dominant element in the American economy. Although they resembled the railroad in scale, organization, and complexity, many relied on new, nonmechanical, forms of power. They included the telegraph and telephone network; the new chemical industry, electric light and power grids; and such linked mass-production-and-use systems as the automobile manufacturing industry (sometimes called the "American" or "Fordist" system), which involved the ancillary production of rubber tires, steel, and glass, and was further linked with the petroleum, highway construction, and trucking industries. In the era when electrical and chemical power was being introduced and these huge systems were replacing discrete artifacts, simple tools or devices as the characteristic material form of the "mechanic arts," that term also was being replaced by a new conception: "technology."[10]

The advent of this typically abstract modern concept coincides with the increasing control of the American economy by the great corporations. In Western capitalist societies, most technological systems (save for state-operated utility and military systems) were the legal property of – were organized as – independently owned corporations for operation within the rules and for the purposes of minority ownership. Thus most of the new technological systems were operated with a view to maximize economic growth as measured by corporate market share and profitability. At the same time each corporation presumably was enhancing the nation's collective wealth and power. Alan Trachtenberg has aptly called this fusion of the nation's technological, economic, and political systems "the incorporation of America."[11] By the late nineteenth century, Thorstein Veblen, an exponent of instrumental rationality, ruefully observed that under the regime of large-scale business enterprise the ostensible values of science-based technology (matter-of-fact rationality, efficiency, productivity, precision, conceptual parsimony) were being sacrificed to those of the minority owners: profitability, the display of conspicuous consumption, leisure-class status, and the building of private fortunes. But the abstract, sociologically and politically neutral (one might say neutered) word "technology" with its tacit claim to being a distinctive, independent mode of thought and practice like "science," conveys no hint of the impress of a particular socio-economic regime.

   Although the English word "technology" (derived from the Greek *techne*, meaning "art" or "craft") had been available, if rarely used, since the seventeenth century, during most of the next two centuries it had referred almost exclusively to technical discourses or treatises.[12] Considering the way historians now routinely (and anachronistically) project the word back into the relatively remote past, it is surprising to discover how recently today's broad sense of "technology" achieved currency: it seldom was used before 1880. Indeed, the founding of the Massachusetts Institute of Technology in 1861 seems to have been a landmark, a halfway station, in its history; however, the *Oxford English Dictionary* cites R. F. Burton's use of "technology" in 1859 to refer to the "practical arts collectively" as the earliest English instance of the inclusive modern usage. It is important to recognize the exact nature of this change: instead of "technology" being used to refer to a written work, e.g., a treatise, *about* the practical arts, it now was used to refer directly *to* the arts (including the actual practice and practitioners) themselves.

   That this broader, modern sense of "technology" was just emerging in the mid-nineteenth century is further indicated by the failure of writers like Karl Marx[13] or Arnold Toynbee, who were deeply concerned about the changes effected by the new machine power, to use the word. At points in his influential lectures on the Industrial Revolution, composed in 1880–81, where "technology" would have been apposite, Toynbee, an economic historian, relied on terms like "mechanical discoveries," "machinery," "mechanical improvements," "mechanical inventions," or "factory system."[14] Yet within twenty years Veblen would be suggesting that the "machine technology" was the distinguishing feature of modernity.[15] My impression is, however, that "technology" in today's singular, inclusive sense did not gain truly wide currency until after World War I, and perhaps not until the Great Depression.

The advent of "technology" as the accepted name for the realm of the instrumental had many ramifications. Its relative abstractness, as compared with "the mechanic arts," had a kind of refining, idealizing, or purifying effect upon our increasingly elaborate contrivances for manipulating the object world, thereby protecting them from Western culture's ancient fear of contamination by physicality and work. An aura of impartial cerebration and rational detachment replaced the sensory associations that formerly had bound the mechanic arts to everyday life: artisanal skills, tools, work, and the egalitarian ethos of the early republic. In recognizing the mastery of various technologies as a legitimate pursuit of the higher learning, the universities ratified the shift from the craft ethos of the mechanic arts to the meritocratic aspirations of the engineering and management professions. The lack of sensuous specificity attached to the noun, "technology," its bloodless generality, along with its habitual use in the more generalized singular form, make the word conducive to a range of reference far beyond that available to the humdrum particularities of the "mechanic" or "industrial" arts. Those concrete categories could not simultaneously represent (as does "technology" itself or, say, "computer technology") a particular kind of device, a specialized form of theoretical knowledge or expertise, a distinctive mental style, and a unique set of skills and practices.[16]

Perhaps the crucial difference is that the concept of "technology," with its wider scope of reference, is less closely identified with, or defined by, its material or artifactual aspect than were the "mechanic arts." This fact comports with the material reality of the new large-scale, complex, technological systems in which the boundary between the intricately interlinked artifactual and other components – conceptual, institutional, human – is blurred and often invisible. When we refer to such systems as compared with, say, carpentry, pottery, glass-making, or machine tool operating, the artifactual aspect is a relatively small part of what comes to mind. By virtue of its relative abstractness and inclusiveness, its capacity to evoke the inextricable interpenetration of, for example, the powers of the computer with the bureaucratic practices within large modern institutions, "technology" (with no specifying adjective) invites endless reification. The concept refers to no specifiable institution, nor does it evoke any distinct associations of place or of persons belonging to any particular nation, ethnic group, race, class, or gender. A common tendency of contemporary discourse, accordingly, is to invest "technology" with a host of metaphysical properties and potencies, thereby making it seem a determinate entity, a disembodied, autonomous, causal agent of social change – of history. Of all its attributes, this hospitality to mystification – to technological determinism – may well be the one that has contributed most to postmodern pessimism.

From the Republican to the Technocratic Idea of Progress

At the time the first complex technological systems were being assembled and the new concept of technology constructed, a related change was occurring within the ideology of progress. It entailed a subtle redescription of the historical role of the practical arts. Originally, as conceived by such exponents of the radical Enlightenment as Turgot and Condorcet, Paine and Priestley, Franklin and Jefferson, innovations in science and the mechanic arts were regarded as necessary but insufficient means of achieving general progress.[17] To the republican revolutionaries of the Enlightenment, especially the radical *philosophes*, science and the practical arts were instruments of political liberation, essential tools for arriving at the ideal goal of progress: a more just, peaceful, less hierarchical, republican society based on the consent of the governed.[18]

The idea of history as a record of progress driven by the application of science-based knowledge was not simply another idea among many. Rather it was a figurative concept lodged at the center of what became, sometime after 1750, the dominant secular world picture of Western culture. That it was no mere rationale for domination by a privileged bourgeoisie is suggested by the fact that it was as fondly embraced by the hostile critics as the ardent exponents of industrial capitalism. Thus Marx and Engels, who developed the most systematic, influential, politically sophisticated critique of that regime, were deeply committed to the idea that history is a record of cumulative progress. In their view, the critical factor in human development, the counterpart in human history of Darwinian natural selection in natural history, is the more or less continuous growth of humanity's productive capacity. But they added a political stipulation, namely that the proletariat's revolutionary seizure of state power would finally be needed if humanity was to realize the universal promise inherent in its growing power over nature. To later followers of Marx and Engels, the most apt name of that power leading to communism, the goal of progress – of history – is "technology."[19]

But the advent of the concept of technology and of the organization of complex technological systems coincided with, and no doubt contributed to, a subtle revision of the ideology of progress. Technology now took on a much grander role in the larger historical scheme – grander, that is, than the role originally assigned to the practical arts. To leaders of the radical Enlightenment like Jefferson and Franklin, the chief value of those arts was in providing the material means of accomplishing what really mattered: the building of a just, republican society. After the successful bourgeois revolutions, however, many citizens, especially the merchants, industrialists, and other relatively privileged (white male) people, took the new society's ability to reach that political goal for granted. They assumed, not implausibly from their vantage points, that the goal already was within relatively easy reach. What now was important, especially from an entrepreneurial standpoint, was perfecting the means. But the growing scope and integration of the new systems made it increasingly difficult to distinguish between the material (artifactual or technical) and the other organizational (managerial or financial) components of "technology." At this time, accordingly, the simple republican formula for generating progress by directly improved technical means to societal ends was replaced by, or rather was imperceptibly transformed into, a quite different technocratic commitment to improving "technology" as the basis and the measure of – as all but constituting – the progress of society. This technocratic idea of progress may be seen as the culminating expression of the optimistic, universalist aspirations of Enlightenment rationalism. But it tacitly replaced political aspirations with technical innovation as a primary agent of change, thereby preparing the way for an increasingly pessimistic sense of the technological determination of history.

The cultural modernism of the West in the early twentieth century was permeated by this technocratic spirit. (A distinctive feature of the technocratic mentality is its seemingly boundless, unrestricted scope, a tendency to break through the presumed boundaries of the instrumental and to dominate any kind of practice.) This technocratic spirit was made manifest, for example, in the application of the principles of instrumental rationality, efficiency, order, and control to the behavior of industrial workers. As set forth in the early twentieth-century theories of Taylorism and Fordism, the standards of efficiency devised for the functioning of parts within machines were applied to the movements of workers in the new large-scale factory system. The technocratic spirit also was carried into the "fine" arts by avant-garde practitioners of various radically innovative styles associated with early modernism. The credo of the Italian Futurists; the vogue of geometric abstractionism exemplified by the work of Mondrian and exponents of "Machine Art"; the doctrines of the Precisionists and Constructivists; the celebration of technological functionalism in architecture by Le Corbusier, Mies Van der Rohe, and other exponents of the international style – all these tendencies exemplified the permeation of the culture of modernity by a kind of technocratic utopianism.

Architecture, with its distinctive merging of the aesthetic and the practical, provides a particularly compelling insight into the modern marriage of culture and technology. Thus the international style featured the use, as building materials, of such unique products of advanced technologies as steel, glass, and reinforced concrete; new technologies also made it possible to construct stripped-down, spare buildings whose functioning depended on still other innovative devices like the elevator, the subway system, and air conditioning. This minimalist, functional style of architecture anticipated many features of what probably is the quintessential fantasy of a technocratic paradise: the popular science-fiction vision of life in a spaceship far from planet Earth, where recycling eliminates all dependence on organic process, and the self-contained environment is completely under human control.

Postmodern Pessimism

But to return now to our initial question: how to understand the current surge of technological pessimism. One way to account for this collective despondency, as I have suggested, is to chart the advent of a crucial word, the abstract noun that names a quintessentially modern class: technology. The point is that the idea of a class called technology, in its ideological inheritance from the practical arts, was suffused from its inception by the extravagant universalist social hopes of the Enlightenment. Those hopes were grounded in what postmodernist skeptics like to call foundationalism: a faith in the human capacity to gain access to a permanent, timeless foundation for objective, context-free, certain knowledge. The stunning advances of Western science and the practical arts seemingly confirmed that epistemological faith, and with it a corresponding belief that henceforth the course of history necessarily would lead to enhanced human well-being.

In their euphoric embrace of that faith, the utopian thinkers of the Enlightenment invented an historical romance called Progress. In it they assigned a heroic role to the mechanic arts. That role, like the romance as a whole, rested on the old foundationalist faith in the capacity of scientific rationalism to yield incontrovertible knowledge. But the part assigned to the mechanic arts in those early years, though heroic, actually was modest compared with what it became after it had been renamed "technology." By the 1920s "technology," no longer confined to its limited role as a mere practical means in the service of political ends, was becoming a flamboyant, overwhelming presence. In many modernist, technocratic interpretations of the romance, "technology" so dominated the action as to put most other players in the wings; in the final act, a happy ending confirmed the vaunted power of technology to realize the dream of Progress. In the aftermath of World War II, however, what had been a dissident minority's disenchantment with this overreaching hero spread to large segments of the population. As the visible effects of technology became more dubious, modernism lost its verve, and people found the romance less and less appealing. After the Vietnam era, the ruling theme of Progress came to seem too fantastic, and admires of the old Enlightenment romance now were drawn to a new kind of postmodern tragi-comedy.

Postmodernism is the name given to a sensibility, style, or amorphous viewpoint – a collective mood – made manifest in the early 1970s. As the name suggests, one of its initial motives was a repudiation of the earlier, modernist style in the arts, and a consequent effort to define – and to become – its successor. The successionist aspect of postmodernism was made clear early on by a series of sudden, sharp attacks on modern architecture, probably the most widely recognized of all styles of aesthetic modernism. As early as 1962, in his seminal essay, "The Case Against 'Modern Architecture'," Lewis Mumford, a leading architectural critic and exponent of early modernism, anticipated many themes of that postmodernist reaction. Most significant was his analysis of the sources of modernism, an architectural style he traced to a set of preconceptions about the historic role of technology. Chief among them, he wrote, was "the belief in mechanical progress as an end in itself," a belief that rested on the assumption that human improvement would occur "almost automatically" if we simply devote all of our energies to science and technology.[20] As in most of his work, Mumford's disapproval was not directed at technology in any narrow or intrinsic sense, not at the mere technical or artifactual aspect of modernity, but rather at the larger ideological context; he was attacking the imperial domination of architectural practice by the overreaching of the technocratic mentality, whereby the technical means, under the guise of a functionalist style, had become indistinguishable from – in fact determined – all other aspects of building practice. His target, in sum, was the dominating role of the instrumental in the later, technocratic version of the progressive ideology characteristic of the era of corporate capitalism.

In making this argument, Mumford allied himself with a dissident minority of writers, artists, and intellectuals who had opposed the technocratic idea of progress for a long time. They were adherents, indeed, of a continuously critical, intermittently powerful, adversary culture that can be traced back to the "romantic reaction" against the eighteenth-century scientific and industrial revolutions. But the cultural dissidents did not abandon the Enlightenment commitment to the practical arts; they rejected the skewed technocratic interpretation of that commitment. What they, like Mumford, found most objectionable was the tendency to bypass moral and political goals by treating advances in the technical means as ends in themselves. Nowhere had this criticism been made with greater precision, economy, or wit than in Henry Thoreau's redescription of the era's boasted modern improvements: "They are but improved means," he wrote, "to an unimproved end." So, too, Herman Melville identified a deep psychic root of this warped outlook when he allowed Ahab, the technically competent but morally incapacitated captain of the Pequod, a stunning insight into his own pathological behavior: "Now, in his heart, Ahab had some glimpse of this, namely: all my means are sane, my motive and my object mad."[21] As the history of the twentieth century has confirmed, high technical skills may serve to mask, or to displace attention from, the choice of ends. And if, as in Ahab's case, the ends are deformed, amoral, and irrational, such a disjunction of means and ends becomes particularly risky.

This kind of flawed technocratic mentality later became a major target of the radical Movement and counterculture of the 1960s. In retrospect, indeed, that astonishing burst of political outrage looks like a last gasp of Enlightenment idealism. It was an attempt to put technology back into the service of moral and political ends. It is important, if we are to understand the genesis of postmodernism, to recognize that it appeared immediately after the events of May 1968, just as the ardent cultural radicalism of the Vietnam era was collapsing in frustration and disillusionment. Thus postmodernism embodied, from its birth, a strong current of technological pessimism.[21] It was a pessimism whose distinctive tenor derived from the adversary culture's inability, for all its astonishing success in mobilizing the protest movements of the 1960s, to define and sustain an effective antitechnocratic program of political action.

In conclusion, let me suggest two ways of looking at the technological aspect of postmodern pessimism. For those who continue to adhere to the promise of Enlightenment rationalism, the postmodernist repudiation of that optimistic philosophy is bound to seem pessimistic. Postmodernism not only rejects the romance of Progress, it rejects all meta-narratives that ostensibly embody sweeping interpretations of history. For those who are drawn to the philosophic skepticism of the postmodernists, however, the repudiation of some of the political hopes that ultimately rested on foundationalist metaphysical assumptions need not be taken as wholly pessimistic. Although such a repudiation surely entails a diminished sense of human possibilities, the replacement of the impossibly extravagant hopes that had for so long been attached to the idea of "technology" by more plausible, realistic aspirations may in the long run be cause for optimism.

But the second way of looking at the role of technology in postmodernist thinking is much less encouraging. What many postmodernist theorists often propose in rejecting the old illusion of historical progress, is a redescription of social reality that proves to be even more technocratic than the distorted Enlightenment ideology they reject. Much early postmodernist theorizing took off from a host of speculative notions about the appearance of a wholly unprecedented kind of society variously called post-Enlightenment, post-Marxist, postindustrial, or posthistoric. A common feature of these theories, and of the umbrella concept of postmodernism, is the decisive role accorded to the new electronic communications technologies. The information or knowledge these technologies are able to generate and to disseminate is said to constitute a distinctively postmodern, increasingly dominant, form of capital, a "force of production," and in effect a new, dematerialized kind of power. This allegedly is the age of knowledge-based economies.

There are strikingly close affinities between the bold new conceptions of power favored by influential postmodern theorists – I am thinking of Jean-Francois Lyotard and Michel Foucault – and the functioning of large technological systems.[23] Thus power, as defined by these theories, is dynamic, fluid, always being moved, exchanged, transferred; it flows endlessly through the society and culture the way blood flows through a circulatory system, or information through a communications network. Unlike the old notion of entrenched power that can be attacked, removed, or replaced, postmodernists envisage forms of power that have no central, single, fixed, discernible, controllable locus. This kind of power is everywhere but concentrated nowhere. It typically develops from below, at the local level, rather than by diffusion from centralized places on high. The best way to understand it, then, is by an ascending analysis that initially focuses on its micro, or capillary, manifestations. The most compelling analogue of the entire system, however, is with the forthcoming mode of fiber-optic communications, an electronic grid that is expected to link all telephonic, television, and computer transmission and reception, and all major data banks, in a single national and eventually global network.

This postmodernism outlook in effect ratifies the idea of the domination of life by large technological systems. It ratifies it by default if not by design. The accompanying mood varies from a sense of pleasurably self-abnegating acquiescence in the inevitable to melancholy resignation or fatalism. In any event, it reflects a further increase in the difficulty, noted earlier, in discerning the boundary between what formerly had been thought to constitute "technology," the material or artifactual armature, which now may be a network of delicate filaments, and the other socio-economic and cultural components of these large complex systems. In many respects postmodernism seems to be a perpetuation of – and an acquiescence in – the continuous aggrandizement of "technology" in its modern, institutionalized, systemic guises. In their hostility to ideologies and collective belief systems, moreover, many postmodernist thinkers relinquish all old-fashioned notions of putting the new systems into the service of a larger political vision of human possibilities.[24] In their view, such visions are inherently dangerous, proto-totalitarian, and to be avoided at all costs. The pessimistic tenor of postmodernism follows from this inevitably shrunken sense of human agency – of irresistible technological determinism. If we entertain the vision of a postmodern society dominated by immense, overlapping, quasi-autonomous technological systems; if the society must somehow integrate the operation of those systems, becoming in the process a metasystem of systems upon whose continuing ability to function our lives depend, then the source of postmodern technological pessimism seems evident. It is a fatalistic pessimism, an ambivalent tribute to the determinative power of technology. But again, the technology in question is so deeply and seamlessly interwoven with other aspects of society that it is all but indistinguishable from them. Under the circumstances, it might be well to acknowledge how consoling it is to attribute our pessimism to the workings of so elusive an agent of change.

Notes

1. Thorstein Veblen, *The Theory of Business Enterprise* (New York: Scribner's 1904, rpt. Mentor, 1932), p. 144.
2. I do not recall ever before having seen the term in print, but I did contribute a paper, "American Literary Culture and the Fatalistic View of Technology," to a conference on "Technology and Pessimism," sponsored by the College of Engineering, University of Michigan, in 1979. See Leo Marx, *The Pilot and the Passenger, Essays on Literature, Technology, and Culture in the United States* (New York: Oxford University Press, 1988), pp. 179–207.
3. Thus Diderot's *Encyclopedia*, a work that epitomizes Enlightenment wisdom and optimism, is a virtual handbook of technologies, most of them of modern origin.
4. In the United States politicians like to call the progressive agenda "the American dream," and it is worth noting that in the recent (1992) election campaign a stock argument of the Democrats was that the current generation may well be "the first whose children are going to be less well off than themselves."
5. As Raymond Williams famously discovered, this dilemma invariably affects efforts to interpret the cultural transformation bound up with the onset of urban industrial capitalism. His own study turned on five key words ("culture," "industry," "class," "art," and "democracy") whose modern meanings and currency derived from the very historical developments he was interpreting. This is of course the historical basis for the "hermeneutical circle" which some regard as vitiating all research in the humanities. See Raymond Williams, *Culture and Society* (New York: Columbia University Press, 1960), preface.

6. To be sure, the prehistory of all words, perhaps especially all nouns, is elusive, for the investigator must devise ways of referring to that for which adequate names were conspicuously lacking. We need a comprehensive history of the word "technology," a project that is, or should be, of primary concern to practitioners of the relatively new, specialized branch of historical studies, the history of technology.

7. Thomas Carlyle, *Critical and Miscellaneous Essays* (New York: Bedford, Clark & Co., n.d.), Vol. III, pp. 5–30.

8. For the modern concept, see Jacques Ellul, *The Technological System*, trans. Joachim Neugroschel (New York: Continuum, 1980); Wiebe E. Bijker, Thomas P. Hughes, and Trevor Pinch (eds.), *The Social Construction of Technological Systems* (Cambridge: MIT Press, 1989); for a recent application of the concept to American history, see Thomas P. Hughes, *American Genesis: A Century of Invention and Technological Enthusiasm* (New York: Viking, 1989). But some earlier social theorists who did not use the same term nonetheless anticipated most features of the concept. Few nineteenth-century thinkers devoted more attention to what we call "technology" than Karl Marx, but though he described industrial machinery as embedded in the social relations and the economic organization of an economy dominated by the flow of capital, he still relied, as late as the first (1867) edition of *Capital*, on "machinery," "factory mechanism," and other relics of the old mechanistic lexicon. "In manufacture the workmen are parts of a living mechanism. In the factory we have a lifeless mechanism independent of the workman, who becomes its mere living appendage." (Robert C. Tucker (ed.), *The Marx–Engels Reader* (New York: W. W. Norton, 1972), pp. 296–297.)

9. Rosalind Williams, in "Cultural Origins and Environmental Implications of Large Technological Systems," *Science in Context* (Fall 1993, in press), argues for a much earlier origin for these systems. She traces their genesis to eighteenth-century Enlightenment philosophers like Turgot and Condorcet, who were committed to the "ideology of circulation." They identified the Enlightenment with the systemic diffusion of ideas and objects in space: ideas through global systems of communication, and objects by means of transportation (road) grids, for the circulation of people and goods. What is not clear, however, is the extent to which circulatory systems of this kind are to be thought of as specifically modern, specifically technological, innovations. After all, the Romans built similarly complex transportation and communication networks. If the point merely is that eighteenth-century theories about circulatory systems anticipated some features – especially the systemic character – of modern technologies, the argument is persuasive. But the systems described in these theories existed primarily in conceptual form, and since they involved no significant material or artifactual innovations, it seems misleading to think of them as innovative "technological systems" like the railroad. A system is "technological," in my view, only if it includes a significant material or artifactual component. Michel Foucault, who first called attention to these theories of circulation, perhaps initiated this idealist mode of interpreting their significance.

10. In explaining the origin of the modern style of corporate management, Alfred D. Chandler describes it as having been "demanded" by the "technological" character of the railroad system. Mechanical complexity, and the consequent need for immense capital investment, were key factors in calling forth a new kind of organization and management. What requires emphasis here, however, is that the effective agent of change in Chandler's widely accepted analysis, the chief cause of the shift, as representative of the technical, from single artifact to system, is the radically new material, or artifactual, character of the railroad. See Alfred D. Chandler, *The Visible Hand: The Managerial Revolution in American Business* (Cambridge, Mass.: Harvard University Press, 1977), p. 87ff, and passim. But historians differ in their accounts of the genesis of the new systems. Thus Thomas Hughes, in *American Genesis* (p. 184ff.),

emphasizes a material or artifactual change, especially from mechanical to electrical and chemical processes, whereas Chandler (whose exemplar, the railroad, belongs to the earlier mechanical phase) emphasizes changes in modes of organization and management.

11. Alan Trachtenberg, *The Incorporation of America* (New York: Hill and Wang, 1982).

12. The OED gives 1615 as the date of the word's first use in English, meaning a discourse or treatise on an art or arts, the scientific study of the practical or industrial arts. See "Technology," *The Compact Oxford English Dictionary*, 1971 ed., Vol. 2, p. 3248; a Harvard professor, Jacob Bigelow, has been credited with anticipating the modern meaning of the word in his 1829 book, *Elements of Technology, Taken Chiefly from a Course of Lectures . . . on the Application of the Sciences to the Useful Arts*. See Dirk J. Struik, *Yankee Science in the Making* (Boston: Little, Brown & Co., 1948), pp. 169–170. Although the scope of the word's meaning has expanded steadily, its current meaning retains its essential etymological links to the practical arts and the material world.

13. On Marx's technological vocabulary, see note 8 above.

14. Arnold Toynbee, *The Industrial Revolution* (Boston: The Beacon Press, 1960), especially pp. 63–66. Toynbee was not timid about using new or unconventional terminology, and indeed these lectures were extremely influential in giving currency to the still novel concept of an "industrial revolution." As late as the 11th (1911) edition, it might be noted, the *Encyclopaedia Britannica*, which contained no separate entry on "technology" in general, was still offering "technology" as a possible alternative to the preferred "technical" in the entry for "Technical Education." See *The Encyclopaedia Britannica*, 11th ed., Vol. 26, p. 487.

15. See note 1 above. Among other obvious indications that the concept was then in its early, avant-garde state of usage would be the way unconventional writers and artists of the period like Herman Melville, Henry Adams, and Oswald Spengler, or the Italian Futurists and Cubists, or (in the next generation) Hart Crane and Charlie Chaplin, used technological images to characterize the distinctiveness of modernity.

16. It is instructive to notice how few of the commonplace statements made nowadays about the import of "technology" in general actually are applicable to the entire range of existing technologies, medical, military, electronic, domestic, biogenetic, contraceptive, etc.

17. Notice the two distinct uses of "progress" here, each with a markedly different scope of reference: (1) the bounded, internal, verifiable, kind of improvement achievable within a particular practice, such as progress in mathematics, physics, medicine, overland transportation, or textile production; the cumulative effect of such manifold kinds of progress doubtless created the conditions for using the word with a much larger scope of reference; (2) a general improvement in the conditions of life for all of humanity, hence a presumed attribute of the course of events – of history, itself.

18. I have summarized this unoriginal interpretation in somewhat greater detail in "Does Improved Technology Mean Progress?" *Technology Review* (January 1987), pp. 33–71.

19. G. M. Cohen, in *Karl Marx's Theory of History* (Oxford: Clarendon Press, 1978), makes a strong case for the view that Marx's conception of history was essentially a version of technological determinism.

20. Donald L. Miller (ed.), *The Lewis Mumford Reader* (New York: Pantheon Books, 1986), p. 75. The essay first appeared in *Architectural Record* 131, no. 4 (April 1962), pp. 155–162. The essay anticipated Robert Venturi's influential 1966 manifesto for postmodernism, *Complexity and Contradiction in Architecture* (Garden City, N.J.: Doubleday [Museum of Art], 1966).

21. See *Walden*, ch. 1, and *Moby Dick*, ch. 41; Henry Thoreau, *The Writings of Thoreau* (New York: Modern Library, 1949), p. 49; Herman Melville, *Moby Dick* (New York: W. W. Norton, 1967), p. 161.

22. American postmodernism is, in my view, most persuasively and attractively represented in the

work of Richard Rorty, but it too is more compelling in its skeptical critique of the philosophical mainstream than in its murky antirealist epistemology. See, especially, *Consequences of Pragmatism* (Minneapolis: University of Minnesota Press, 1982), and *Contingency, Irony, and Solidarity* (Cambridge: Cambridge University Press, 1989). For an acute assessment of the political weaknesses inherent in this outlook, see Christopher Norris, *What's Wrong with Postmodernism: Critical Theory and the Ends of Philosophy* (Baltimore: Johns Hopkins Press, 1990). There have been many efforts to define postmodernism, but perhaps the anti-Enlightenment aspect important here is most clearly set forth by David Harvey, *The Condition of Postmodernity* (Oxford: Basil Blackwell, Ltd., 1980), and Jean-Francois Lyotard, *The Postmodern Condition* (Manchester, 1984).

23. Lyotard, *The Postmodern Condition*; idem, *Le Differends* (Minneapolis: University of Minnesota Press, 1984); Michel Foucault, *Power/Knowledge* (New York; 1972).

24. This is not true of all postmodern theorists; thus Richard Rorty reaffirms a traditional liberal perspective, though one whose capacity to provide a theoretical basis for the control of technologically sophisticated multinational corporations is extremely uncertain.

TECHNOLOGY AND THE ILLUSION OF THE ESCAPE FROM POLITICS

YARON EZRAHI

Hebrew University of Jerusalem

I would like to advance the argument that during the closing decades of the twentieth century a growing public distrust of science and technology is more a symptom of a transformation of the conception of politics than of changes in science and technology. Yet even if this transformation in the conception of politics is not much affected by science and technology, it has in turn radical consequences for the status and role of science and technology in modern democratic states.

Central to the early modern conception of politics was the idea that technological action is apolitical action, and that when it is enlisted in the service of government it can constitute a possible mode of depoliticizing the use of political power. This sharp distinction between technological and political actions was based on the premise that nature and society are two separate and distinct domains.

It is on the basis of such a distinction that the mastery of nature – to which Robert Pippin has referred in his paper in this volume – could appear as a universal human goal, as something common to all members of human society, thus removing from the idea of technology the problem of harsh normative choices.

Before it was, at least partly, undermined by tendencies to socially relativize or to politicize conceptions of nature in the late twentieth-century liberal-democratic polity, the force of this cultural formula was so great that it could invest technological means with sufficient symbolic power to render the goals they seemed to serve apolitical or incontestable. That made technology one of the most powerful rhetorical means for generating political consensus. Viewing physical reality as an external constraint, something that resists the human will, served for a long time to justify any technical device that could control or utilize nature as extra-political, a progress for "humankind" – not just for one or a few groups at the expense of others.

The separation of nature and society, or the distinction between technology and politics, sustained one of the most central and probably most useful elements of the Enlightenment liberal-democratic vision: faith in the possibility of escape from politics, from the underdeterminism of human judgment to mechanical determinism, from the anxious openness of human choices to the technical closure of computation – an escape from the need to rely on elusive and fragile human judgments or on the moral character of human actors to a less precarious trust in experts and their technical competence.

Since liberal democracy has been based on the reassertion of the role of individuals in politics, the fear of the overpersonalization or subjectivization of the exercise of political power, of the capricious uses of authority, has led to a constant preoccupation with democratically legitimate modes of depersonalizing the exercise of power: *technical discipline* appeared often as a more publicly reliable constraint than *moral discipline*.

Just as the act of a parliamentary majority (which is, of course, a political act) can produce a law that in turn can be used to set limits on partisan politics, so technological action can be apolitical only at a derivative, second-order level. Its status as apolitical presupposes a prior, and politically loaded, if latent, attitude that considers actions directed at controlling or manipulating nature as qualitatively different and separate from actions directed at affecting society.

In the Western cultural tradition the idea of nature as external and indifferent to humanity depend on a host of complex theological and metaphysical developments, which were congenial to the rise of the modern scientific view of nature as mechanism and to the predisposition to distinguish moral-religious from instrumental-technological types of action.[1] This view of nature played a central role in the consolidation of modern conceptions of the human actor as an agency in the field of action and of his position in relation to history and technology.[2]

It is the willingness to take these fundamental conceptions for granted that has been revoked toward the end of the twentieth century as the word "nature" has been increasingly replaced by other words such as "environment" and "ecology." These words refer to relations or interactions between "nature" and "society," or "nature" and "humans," rather than to a domain named "nature" which is distinct and separate from a domain called "society." In the late twentieth century it has become more widely acknowledged that conceptions of nature are embedded in wider orientations towards the relations between society and the physical world, that there is no conception of nature that is not, in some sense, a form of what Emile Durkheim called "collective representation," the cultural product of a particular society.[3] The increasing unwillingness to accept the earlier conception of nature as universally external to human society or as a neutral, extrasocial, or extrapolitical referent of human action has been

undermining the cultural foundations of established technological and professional authorities with which the Enlightenment liberal-democratic vision, at
least partly, replaced the traditional religious and moral bases of public action.
The shift from moral to instrumental paradigms of action, from a focus on
self-mastery to a focus on the mastery of nature, represents, therefore, a major
change in Western cultural orientations. Within the more particular context of
modern liberal-democratic politics this shift has been connected with the tendency to become less preoccupied with the inner, invisible relations between
the individual and his or her soul or conscience, and more with the external,
transparent relations between the means and ends of human actions. While the
drama of moral struggle is characteristically internal, the drama of technological
action is characteristically external and more publicly visible.

An historian of nineteenth-century American technology observed that in
response to the demand that engineers adopt a code of ethics like lawyers, the
engineers' association felt that "since the actions of the engineers were checked
at every point by the immutable laws of God and nature there was no possibility
of undetected fraud."[4] The idea that an engineer is impervious to corruption
because of the inescapable visibility of his actions seemed to realize the kind of
transparency that Jean Jacques Rousseau despaired of achieving in the moral
sphere.[5] It is because of such pessimism about unaided moral politics that
modern liberal-democratic ideology invested so much in legalism as well as
in technological instrumentalism as external referents of action. Both legalism
and technicalism appeared to be promising substitutes for moral virtue and
honesty as reliable frameworks for judging actors, or as the sources of external
standards of adequate and accountable political action. Philip Fisher sees a link
between transparency and actions mediated by technology. "Because there is a
negotiation built into all objects and tools – all output – between the facts of
the natural world and the facts of the body, there is transparency." He contrasts
the transparency of work and output in an industrial society with the opacity of
ritual and sees a society that is manifest in economy rather than in religion as a
society capable of giving rise to "democratic social space." [6]

The trend to define action in instrumental technical terms was responsive
during the nineteenth century to the fear that directly spiritualizing or mor-
alizing politics leads to unresolvable conflicts. Following the move of liberal
thinkers like Locke to place politics in the realm of materially tangible things, it
was hoped that at least some disputes could be resolved by reference to facts.[7]
What was consistently ignored for a long time is that any technology, and indeed any technical action, has a hidden ethical and political "software"; that
the transparency of actions captured within ends-means schemata conceals a

multiplicity of normative choices that do not fully lend themselves to technical or computational treatment. Advocates of science- or technology-based democratic politics overlooked the fact that engineers do not just solve technical problems. They also generate changes in the distribution of values and scarce material and political assets.

The hidden normative or ethical software of machines and techniques was exposed, however, as soon as the process of democratization revealed that when the citizens are free to express their preferences, the result is not the convergence of goals but the articulation of their irreducible heterogeneity. In late twentieth-century democracies the equation of individuality with uniqueness and, beyond that, the democratization of uniqueness as a possibility that is not confined only to the privileged classes but extends to all the citizens, has undermined the idea of majority as a near-mechanical aggregate of commensurable individual choices. Without this idea, neither society nor the state could appear as unproblematic, transpersonal superagencies of technical collective actions. During the second half of the twentieth century, the tendency to supplement economic perspectives on the "revealed preferences" of individuals with social indicators and the analysis of political and ethical choices have been among the symptoms of this change.[8]

The political mandates that a society of unique individuals (as well as unique groups) could produce rarely support large-scale technological actions by the state. Instead the policies and actions of modern democratic governments appear increasingly to be the results of necessarily eclectic, fuzzy and shifty compromises.[9] In the context of such political compromises, monumental technological projects simply do not make sense. The modern preoccupation with side effects, spillover effects, and the externalities of technological developments is not so much a reflection of a growing knowledge of the impact of science and technology on society, the environment, etc. (although there is, of course, some of that). It is, in my opinion, rather a necessary result of the discovery that the normative basis of public opinion and of public action is inherently heterogeneous, diverse, and internally contradictory; that public values cannot be unproblematically deductible from private values, that private values are often internally incompatible, and that public values are often incommensurable. It reflects the crystallization of social attitudes which no longer accept democratization as the massive application of political power for the advancement of agreed-upon, clear public goals. Instead, these attitudes increasingly equate democratization with the disempowerment of the national government, with dividing its powers among regions, institutions, and issues.[10] In a political context in which there is a strong impulse to decentralize the normative mandates and results of public action, to spread out the costs and benefits of collective actions across a wide spectrum of values and interests, no action can be perceived as inherently apolitical, as strictly technical and therefore in this sense neutral in relation to all individuals or groups simultaneously. The application of the optimal means to achieve the highest goal of one party may very well be the imposition of a low-priority goal on another.

Hence late twentieth-century signs of technological pessimism can be at least partly construed as the result of the impossibility of operationalizing the concept of "the public interest" rather than of the unanticipated side effects of technological change. The political dimension of technology is not a property of technology proper but of the relations between certain aspects of technology and certain aspects of the socio-political context that are engaged when a particular technology is applied. The "political relativity" of technology does not presuppose, therefore, technical inadequacies. It is inherent in the interaction between technology and the legitimately heterogeneous (and often internally inconsistent) normative terms of action in the public sphere.

Recognition of the importance of this political fact is probably the reason behind Leo Marx's observation (in this volume) that the impact of technology on nature has been a relatively recent concern. Technological pessimism is a symptom not of the "decline of technology" as a body of applied knowledge per se but of the erosion and fragmentation of mandates for collective techno-logical action in the political domain. It often represents the recognition that since technology is not simply an embodiment of scientific knowledge but also a branch of ethics and politics, it does not have the authority or the symbolic power to limit or replace ethical discipline and political choices. In the final analysis, then, *technology does not provide an escape route from politics, and the arbitrary use of political power cannot be checked by nonarbitrary technical norms.* If the European Enlightenment produced the hope that because arbitrariness is an irrational use of freedom its opposite is the rational use of freedom, which is compatible with order and manifest in technology, in late twentieth-century Western society we have come to recognize freedom as something manifest not so much in rational systematic actions as in benign disorder, not in planned systems, which seem to defy capricious and disorderly actions, but in eclectic actions and fuzzy, shifty compromises, which reveal the absence of a controlling agent. As such, what I have called technological pessimism is also an aspect of the blurring of the earlier demarcation lines between the private and the public spheres.

To some, the loss of meliorist-universalistic technological public action that can shape the future is bound to appear regressive. To others, however, the destruction of the possibility of public action by the decentralization of the power to authoritatively define the purposes and terms of political action is a positive development, and technological pessimism, as the recognition of the limits of technology, is a progressive shift. A growing understanding of the existence of irreducible value conflicts at the heart of democratic politics inevitably discredits also what can be identified as a "solution" approach to policy problems. The solution approach presupposes that social problems can be defined clearly and compellingly, and that technical experts and scientific professionals can then advise policy makers on which are the optimal means to solve such problems. But in late twentieth-century democracies, it is usually naive to expect an agreement on the definition of a social problem and on the means to solve it. Instead, one more typically confronts a situation where competing political and economic interests generate competing definitions of the problem (including those who say there is no problem at all). The disagreement about how to define the problem inevitably extends to attitudes towards the means to be applied. Modern policy analysts and scientific advisers to democratic governments are increasingly persuaded that in the context of normative and political pluralism, the solution approach to social problems is anachronistic, ineffective, and illegitimate.[11]

Once pluralism, or the irreducible conflict of values and interests, is accepted as a given, the problem of relating knowledge to legitimate power, of combining science and policy decisions, is no longer conceived as the problem of how to replace irrational political considerations with rational scientific ones. The problem could be recast rather as how to find the best scientific and technical implementation of decisions based on political compromise.

The question, therefore, is not how to substitute scientific solutions for political decisions but how to make such inevitably political decisions better from a scientific and technical point of view. Many experts wonder whether the political compromises that produce the decisions of democratic governments can be improved by scientific knowledge and technical skills and how the eclectic and often fuzzy decisions that are produced by a balance of power among government officials, private economic interests, workers, citizens' groups and representatives of other publics can be scientifically and technically informed and improved.

A growing number of experts point out that political compromises typically mirror power relations, not relations among relevant ideas or arguments; that political compromises do not eliminate contradictions and tensions among alternative positions but preserve them; and that public policies produced by modern democratic governments are typically eclectic, self-contradictory, and hence indifferent to the tests of logical consistency and instrumental-technical rationality. How can such policies, they ask, be improved by scientific and technical knowledge? Moreover, they say, whereas, for scientists and experts, decisions are usually directed at solving problems, for politicians, decisions are characteristically the means to legitimate their authority and their power as policy-makers. In such context the advice of scientists and experts tends only to embarrass the politicians because it exposes the gap between what could be done technically in order to solve a problem and what is politically feasible considering the limits imposed by the logic of compromise and power. Despite the above constraints, the record of advanced Western democracies demonstrates that scientists and technologists can help improve the instrumental adequacy of policies generated by democratic political compromises. Such improvements are manifest in cases where the political needs of the decision-makers rather than being ignored or overlooked are balanced against the functional or technical parameters of policy decisions and actions, where political compromises are formed against a background of reliable information on what is likely to work. Such balances or compromises are more likely to be achieved when the following conditions are satisfied:

i.   When some usually more academically oriented or independent experts assume a more comprehensive public perspective and define the reasonable technical limits of legitimate political compromises and policies with an eye to the wider collective interests of the actors in the political arena. Such experts are likely to be more effective if, instead of attempting to identify or advance one optimal, ideal technical solution, they were to present a range of reasonable technical solutions which could be taken into consideration in the course of the process of putting together the necessary political coalitions.

ii. When the competing groups or interests employ experts who can translate their partisan perspectives into technically relevant arguments and measures. In such cases the experts who work, for example, for trade-unions, business firms, citizens' groups, environmentalists, etc. help shift the focus of the interaction among the parties from the articulation of competing interests and principles to the design of balanced yet working solutions.

iii. When the more public oriented experts succeed in projecting in the political context of public policy-making the implications of alternative decisions or policies *over the longer term spans* in addition to short term projections in relation to the present-oriented concerns of the political and economic interests.

iv.   When experts can identify and articulate short-term economic, social, and political incentives for taking actions that are instrumental to the advancement of public values in the long term. Often when governments cannot be persuaded to take the "right" actions for the "right" reasons, that is functionally or technically valid ones, they are persuaded to take such actions for "wrong" reasons. This state of affairs exists in every democracy because of the inevitable discrepancy between conditions that render decisions politically legitimate and conditions that make decisions technically or functionally rational.

This is also, of course, an important symptom of the discontinuities between the perspectives of politicians and technical experts on policy problems. Ultimately, of course, they must find ways to cooperate because in the long run political decisions unguided by analysis are blind – just as analysis insulated from public affairs is irrelevant. Modern democratic publics complicate the lives of both politicians and experts, but more than anything else they penalize them for unwise decisions.

Notes

1. See for instance, Sheldon Wolin, "Post-Modern Politics and the Absence of Myth," *Social Research* 52 (1985), 217–239. See also Frank E. Manuel, *The Religion of Isaac Newton* (Oxford: The Clarendon Press, 1974).
2. See Yaron Ezrahi, *The Descent of Icarus: Science and the Transformation of Contemporary Democracy* (Cambridge, Mass.: Harvard University Press, 1990), especially Part II.
3. Emile Durkheim, *The Elementary Forms of the Religious Life* (New York: The Free Press, 1965).
4. Monte A. Clavert, *The Mechanical Engineer in America* (Baltimore: Johns Hopkins University Press, 1962), p. 265.
5. For a brilliant and extensive discussion of Rousseau's idea of transparency, see Jean Starobinski, *J. J. Rousseau, Transparency and Obstruction* (Chicago: The University of Chicago Press, 1988).
6. Philip Fisher, "Democratic Social Space: Whitman, Melville, and the Promise of American Transparency," *Representations* 24 (Fall 1988), 60–101.
7. See in particular J. Locke, *Letter Concerning Toleration* (Indianapolis: Bobbs-Merrill, 1955), pp. 17, 18, 19, 56–69.
8. See the discussion of this point in Ezrahi, *Descent of Icarus*, pp. 263–269.
9. See for instance John Rawls, "Justice as Fairness: Political not Metaphysical," *Philosophy and Public Affairs* 14 (1985), 225, and Yaron Ezrahi, "Utopian and Pragmatic Rationalism: The Political Context of Scientific Advice," *Minerva* 18 (Spring 1980), 111–131.

10. See for instance Samuel P. Huntington, *American Politics: The Promise of Disharmony* (Cambridge, Mass.: Belknap Press, Harvard University Press, 1981).

11. Ezrahi, "Utopian and Pragmatic Rationalism."

JOSEPH GLANVILL'S PLUS ULTRA AND BEYOND: OR HOW TO DELAY THE RISE OF MODERN SCIENCE

KLAUS REICHERT

Johann Wolfgang Goethe-Universität, Frankfurt am Main

With the seventeenth century we associate the rise of the mechanist world picture, which held sway almost unchallenged as the paradigm of scientific thinking up to the threshold of our own century. Seen in retrospect, it looks as if this rise had been straight-lined, incessant, and necessary, the logical outcome of the newly established ideals of method, empiricism, rationality, and noncontradiction. Seen in the context of the Restoration period, we get an altogether different picture, even though the principles of method, empiricism, and rationality were adhered to almost uniformly, however differently interpreted.[1] What we see is an extremely controversial picture, a medley of interlocking political, religious, scientific and literary debates. To extract one strand of thought, to follow closely one argumentative line only – say a scientific one – seems to be next to impossible or almost purely hypothetical. To speak about science means at the same time to become enmeshed in religious strife, to take up sides with or against Hobbesian theory of society, Hobbesian "materialism," etc. – a situation not unfamiliar to us today.

If the arguments shored up against the rising mechanist world picture were discarded by historians of science or philosophy because they had been brought forth not by proper scientists but by theologians, we would first have to pose the question of how to draw the line between "proper" and "improper" scientists. Most of the men who founded the first bastion of science, the Royal Society, or became members of it, were so-called *virtuosi*, men who dabbled in science because science had become fashionable, but apart from that were landed gentlemen, literati, or clergymen of various denominations, mainly Puritan or Anglican. Even so conspicuous an aspirant to the title of a proper scientist as the sceptical chemist Robert Boyle published, alternating with his studies on air, motion, hydrostatics, etc., tracts about *The Excellency of Theology compared with Natural Philosophy* (1673) or *Some Considerations about the Reconcilableness of Reason and Religion* (1675). And it came as a shock to Newton adepts when it was discovered in the 1930s that the prince of physicists had spent half his working time on the study of alchemy and similar subjects.[2]

We have to bear these cross-faculty givens in mind if we are to deal with the complexity of science in that period without losing sight of the issues that were noncontributory to the mainstream of scientific development. We have to ask why these issues were furiously debated at the time and also why they were discarded later on, based on a different notion of reason. We are faced with the paradox that the same people who advocated the new mechanist approach to nature adhered to the notions of a nature inhabited by spirits and governed by occult forces. These were "differences of degree between varying conceptions of nature," as was recently shown by Stuart Clark, not a "difference of kind between the 'scientific' and the 'occult' approach."[3] Two paradigms were at work at the same time and sometimes within the same minds, without any apparent contradiction. The contradictions are to be sought elsewhere – in the assumptions of orthodoxy or heterodoxy as to the status of spirits, the role of revelation, the compatibility of faith and reason. The question that interests us here, then, is not so much the old one of opposing "pre-scientific" and "scientific" modes of thinking. The very intensity of the debates points to the fact that something larger was at stake: the place of science in society. What were the implications if it became "socially neutral,"[4] if it ceased to be anthropocentric or abandoned mutual coherences?

The triumph of the mechanist paradigm was not as certain as it may have appeared to later ages. In what follows I shall focus on one of the attempts to formulate an alternative archetype – or to give to the mechanist paradigm a different bend by having it exclude less – the attempt brought forth by the Cambridge Platonists, Ralph Cudworth (1617–1688), Henry More (1614–1687), and Joseph Glanvill (1636–1680). The interesting point about these Royalist, strictly anti-Puritan thinkers is that they started off as sceptics, became fervent advocates of Descartes, and ended (still sceptics) as his critics because they felt that his dualism – especially if combined with Hobbesian materialism – ultimately would lead to atheism. They remained adherents of the possibilities opened up by the new scientific thinking but refused to dissociate it from religion, ethics, and society in that they pursued the consequences of one for the other. Their arguments may seem to us contradictory and spurious, even paradoxical, but if we are prepared to set them in a broader perspective, to allow for a dialogue of different discourses, to translate their seemingly weird preoccupations back into their rational foundation, we may arrive at a challenge the validity of which we can begin to grasp again today, after the mechanist world picture has come to its end. I am going to examine the strange but instructive case of Joseph Glanvill, who started out as Baconian and ended as a staunch defender of the existence of witches. Was this a backlash into superstitious darkenings or a step forward towards a broader notion of enlightenment?

Thomas Sprat's *History of the Royal Society*, which he presented to his fellow members in 1667, was intended as an apologia of that institution, which had been under constant attack since its founding several years earlier from varied sources: the Royal College of Physicians, the universities, the Church, and the Puritans. Sprat, an ambitious young theologian later to become bishop, was conciliatory in tone, more interested in the elegance of his style than knowledgeable about the radical implications of the reorientation in thinking being pursued by the more scientifically minded members of the Society. Unsatisfied with this attempt, Glanvill was encouraged to write a second apologia, his *Plus Ultra* of 1668. The hero of both works is Francis Bacon, but a change of emphasis is evident. When Sprat singled out one man among the members of the Society as the embodiment of all its endeavours, it was Sir Christopher Wren, geometrician, architect, inventor, technologist, and historian of the weather etc., a latter-day specimen of the *"uomo universale,"* whom Sprat held to be superior to Descartes in his "Doctrine of Motion." For Glanvill – and this puts his account into closer alignment with what the Society was to stand for – the leading figure was Robert Boyle, the chemist.

The very title of Glanvill's book is a reference to Bacon. *Plus Ultra* was Bacon's (as well as Emperor Charles V's) motto, the catchphrase of the great instauration of curiosity: to step beyond, to pursue untrodden paths, to leave the bounds of the known and to sail through the Pillars of Hercules. In his technological utopia *Nova Atlantis*, Bacon had cited the great inventions already made – from the wheel and the plow to the telescope, gunpowder, and the printing press – and had envisaged scores of others to come. Reading Glanvill we get the impression that Bacon's dreams had come true. We get the impression that Glanvill's age was far superior to all past ones by the sheer bulk of what it had achieved. And Glanvill does not confine himself to an enumeration of these achievements – as Bacon seems to do when he exposes them in a kind of museum – for we are made aware of an ongoing process of which we become the onlookers. We breathe the air of incipient progressivist thinking – the future is open or, in Glanvill's words, "Geometry is improving daily." (1668; p. 37) All the new sciences pass muster: chemistry (though not that of Paracelsus who, as a hermeticist, is not mentioned at all; nor is, for that matter, Van Helmont!), anatomy, Harveyan medicine, geography, astronomy. In contrast with Bacon, the importance of mathematics is stressed – "There was a time when these [the mathematical 'arts'] were counted Coniurations" (1668; p. 20) – and in particular the role of the new geometry is praised, for "without it we cannot in any good degree understand the *Artifice* of the *Omniscient Architect* in the composure of the *great* World, and *our selves*. ... the *Universe* must be *known* by the *Art* whereby it was *made*. ... *Geometry* is a most *useful* and *proper help* in the affairs of *Philosophy* and *Life*." (p. 25f.) The name of the great Descartes still much disputed elsewhere, is glorified, for he was "designed by *Heaven* for the *Instruction* of the Learned World ..." (p. 33) and "hath proposed an *Universal Method* for the *Solution* of all *Problems*." (p. 35) Formulations like these suggest that he was not all that familiar with what Descartes had done, nor with Galileo, whom he mentions but then asks, in discussing the telescope, whether posterity may "find a *sure* way to determine those *mighty Questions, Whether the Earth Move?*" (p. 55)

Bacon's philanthropic intention of doing research for the benefit of mankind was still evident in such set phrases as "the philosopher" must work "either for *Light* or *Use*." (p. 52) In fact the stress is clearly laid on light, enlightenment, knowledge. The useful inventions Glanvill parades – telescope, microscope, loadstone, thermometer, barometer, airpump – are but instances of how man's knowledge of nature has been enlarged and carried into its minutest recesses, where human sense faculties hitherto fell short. A piquant example is gunpowder: with its help the people of the Americas were conquered, consequently "*new Plants, new Fruits, new Animals, new Minerals*, and a kind of *other World of Nature*" (p. 73) enriched the realm of known nature. "And so these *Engines* of *Destruction*, in a *sense* too are *Instruments* of *Knowledge*." (p. 82) The primacy of the pursuit of knowledge could hardly be more explicitly stated. At the same time there is an awareness of the "dialectics of Enlightenment."

The proper study of nature is certainly not through logical reasonings, not an "*establishment* of *Theories*, and *Speculative Doctrines*" (p. 89) – here we already sense the implicit anti-Cartesian bias of his arguments – but through observation and experiment: "carefully to *seek* and faithfully to *report* how things are *de facto*." (p. 89) After Glanvill's having maintained the limitations of the senses, it may come as a surprise that in his attacks against speculation and Aristotelianism we find an emphatic vindication of these very senses. The aims of the Royal Society are, he states, "to free *Philosophy* from the vain *Images* and *Compositions* of *Phansie*, by making it *palpable*, and bringing it down to the *plain objects* of the *Senses*; For those are the *Faculties* which they [the members of the Royal Society] employ and appeal to, and complain that *Knowledge* hath too long hover'd in the *clouds* of *Imagination*." (p. 89) Plain objects of the senses? With or without the help of artificial extensions? Where do sense data end; where does speculation being? Dominance of the concrete properties of a "fact" over its classification? Unmediated observation of "naked objects" without a reference frame of theoretical assumptions? We must bear these questions in mind, for at the same time Glanvill is defending the existence of witches on the basis of sense evidence against their [imaginary] denial.

Bacon's novel idea of investigation as a joint venture – "an *Assembly*, that might *intercommunicate* their *Tryals* and *Observations*, that might jointly *work*, and joyntly *consider*" (p. 88) – had found its realization in the efforts of the Royal Society, mainly in the form of communicating, comparing, and sifting material won by way of concrete observation and experimentation; in the words of Boyle, "they are fitted with Opportunities to amass together all the considerable *Notices, Observations*, and *Experiments*, that are scattered up and down in the *wide* world; and so, to make a *Bank* of all the *useful Knowledge* that is among men." (p. 108) This project of a large treasure house of knowledge seems to be more in accord with the novel idea of compiling encyclopedias than with the equally new approach to scientific procedures. Glanvill counts on inclusion; the emerging scientific paradigm is built on exclusion. In a surprising aside Glanvill makes us aware what kinds of materials are to be collected: not only the things obtained by methodically guided research, as the future development of science would demand, but also "those things which have been found out by *illiterate Tradesmen*, or lighted on by *chance*." (p. 105) And further: "by the *Knowledge* and *application* of some *unobvious* and *unheeded Properties* and *laws* of natural things, divers *Effects* may be produced by other *means* and *Instruments* than those one would judge *likely*; and even by *such*, as if proposed, would be thought *unlikely*." (p. 105) It is only now that we can grasp the relevance of statements like these: it is an implicit warning against the "vanity of dogmatizing"[5] inherent in any scientific community; it disregards the limitations of specialization, which were beginning to be defined; and it pleads for the inclusion of seemingly disfunctional, marginal, or external elements that may yet tip the scale. Glanvill's fight against prejudices extends to the possibility that even illiteracy may have something to teach.

In his advocacy of the new scientific paradigm, howsoever enmeshed with pretheoretical modes of thought, Glanvill again and again feels the need to defend himself and the Society against the reproof of impending atheism. The aim is not, he contends, to foster science for science's sake, nor is it to better man's condition – the word "usefulness" seems to be restricted to the coherence of an explanation; the aim is rather "to lay the *Foundations* of *Religion sure* ... indeavouring to secure the *Foundations* of the *Holy Fabrick*." (p. 138f.) The pursuit of this aim has effected numerous impressive results, illuminated many secrets, dissolved superstitious beliefs, and dispelled vain speculations, for "the *Experimental learning* rectifies the *grand abuse* which the *Notional*

Knowledge hath so long foster'd and promoted, to the *hindrance* of *Science*, the *disturbance* of the World, and the prejudice of the *Christian Faith*." (p. 148f.) Yet Glanvill's apologia does not end on an optimistic note; the sceptical or pyrrhonian spirit breaks through. All that man can do is to exclude "the *malign influence* of our *Affections*" (p. 146; this is directed against the Puritans) and to rely only on the power of reason even in things religious, but what does this amount to after all, in the face of our "own Fallibility and Defects"? "I say, the *Free* and *Real* Philosophy makes men deeply sensible of the infirmities of human Intellect, and our manifold hazards of *mistaking*, and so renders them *wary* and *modest*, *diffident* of the *certainty* of their *Conceptions*, and averse to the *boldness* of *peremptory asserting*." (p. 146) What had been undertaken as a vindication of the novel enterprise "for the solution of all Problems" ends by pointing to the doubtfulness and unreliability of the faculties of the human mind, without, however, much to Glanvill's credit, introducing the category of belief.

Glanvill's apologia for the Royal Society is encompassed by his works on witchcraft. Two years before his *Plus Ultra* he published an essay titled *Philosophical Considerations Concerning Witchcraft*. He reissued the essay in 1668,[6] the year of *Plus Ultra*, together with a revamped version of an older essay, *A Blow at Modern Sadducism in Some Philosophical Considerations about Witchcraft*, which in turn led to the monumental volume *Saducismus Triumphatus, or Full and Plain Evidence Concerning Witches and Apparitions* in 1681. The subject of witches, then, is congruent with his advocacy of the new science. Is it a contradiction, a relapse into superstition? Glanvill feels that it is not, and he is backed in this by More and Boyle, two of the most eminent members of the Society. The introduction of witches is meant to point, rather, to some of the shortcomings of the new paradigm. The very term "sadducism" indicates the drift of the argument: in the intertestamental period, Sadducees were members of a "sect" that denied the resurrection of the dead and the existence of angels and spirits; in the seventeenth century the name became a derogatory term for materialists. (The OED cites Baxter: "Hobbists, Infidels, Atheists, Sadduces".)

After all, Glanvill asserted, is it reasonable to deny the existence of spirits on the grounds that we cannot see them when we know that our senses are infirm and unreliable? Have our perceptions not become extended by artificial means such as the telescope and the microscope? Should it not be possible that at some point in the future the vast realm of spirits may become accessible for observation? We have to remember that in those very years Newton presented the results of his famous experiments on light and colors to the Society: that light could not only be split up but that it was corporeal, hence had a substance, and that it consisted of a "heterogeneous mixture of differently refrangible rays." (p. 72) Sent through a prism, light could be made visible in its colored elements, and sent through a second prism it could be made to disappear again. Certainly the existence of witches and spirits could not yet be proved scientifically, but nor could it be denied, for it was tangible in countless manifestations. It was one of many phenomena for which the machinations could not yet be accounted.

We cannot conceive how the *Foetus* is form'd in the Womb; nor as much as how a Plant springs from the Earth we tread on; we know not how our Souls move the Body; nor how these distant and extream Natures are united; ... And if we are ignorant of the most obvious things about us, and the most considerable within our selves, 'tis then no wonder that we know not the constitutions and Powers of the Creatures, to whom we are such strangers. (1675; p. 7)

Again and again Glanvill cites analogies from the nature we know: the biting of a mad dog, the transmission of infections by "certain tenuious Streams through the Air," the appearance of wounds in a body not caused by external weapons but "inflicted by the Imagination," the separation of body and soul as in epilepsy or madness, etc. And why may not afflictions like these in the last resort be attributed to the influence of some malign spirit? All Glanvill can do so far is to collect and compare what he calls "the evidence" in much the same manner he describes in *Plus Ultra*.

Glanvill was, of course, not alone in asserting the existence of witches and apparitions. On the contrary, there were fewer people – "atheists" – who denied their existence than there were vindicators. Stuart Clark has drawn attention to the fact that this is not at odds with Bacon's program since he proposed (in *De augmentis scientiarum*) a natural history of "pretergenerations" – "the Heteroclites or Irregulars of nature".

This proposal certainly made the marvelous a central rather than a peripheral category of investigation. ... Singularities and aberrations in nature were not merely correctives to the partiality of generalizations built on commonplace examples; as deviations from the norm they were especially revealing of nature's ordinary forms and processes.[7]

Indeed, Bacon had argued for the inclusion of "narratives of sorceries, witch-crafts, charms, dreams, divinations, and the like" into the scope of enquiry, "where there is an assurance and clear evidence of the fact," "for the further disclosing of the secrets of nature."[8] Demons were held to work within the boundaries of natural causation – "Nature [is] Master and Commander of him [the devil]," wrote John Cotta in 1616[9] – even though "their causation was obscure and hidden from men."[10] Thus the program adhered to by Glanvill and numerous others was one of radical enlightenment. It was a broad scientific project and should not be confused with what has been termed the European witch-craze. Yet this is only part of the picture and has to do with the epistemological questions of how to define nature and its limits and of how to delineate the field of research.

But why the increasing intensity of the debate; why the raging battles in the name of reason far beyond the borders of dissenting scholarly opinions? Several answers seem plausible. One is that it was to point to problems raised by the mechanist paradigm, e.g., how does the mind move the body if both are as rigidly separate as is supposed, how do even the parts of a body cohere if their union "consist only in Rest; it would seem that a bagg of dust would be as firm a consistence as that of Marble or Adamant"? Or what happens to the soul when it is separated from the body in death? Questions like these point to gaps in the new model, problems unsettled, things deliberately left out. It is the evidence of observed facts, their unity and coherence, their meaningfulness, that Glanvill shores up against a *theoretical* model based on clearly defined and distinct reductions. Glanvill probably did not sense the methodological contradiction, but what he did sense was that the new model excluded the very things he felt to be real and most in need of understanding and verification. This seems to be one of the chief predicaments when a change in paradigm occurs: why doesn't the new one explain as many causes as the former had done? Another reason for Glanvill's vindication was to fight Sadducean or Hobbesian atheism disguised as materialism: "If the Notion of a Spirit be absurd, as is pretended; that of a GOD, and a SOUL distinct from Matter, and Immortal, are likewise Absurdities." (p. 4) Another reason, which would account for the reasonable and "objectivist" method of his proceeding, was to fight the Enthusiasts, Fanatics, or Illuminationists, i.e., certain sects of the Puritans who were guided by their passions and affections and reduced witches to individual insanity, personal sin, or socially caused adversity.

It is not difficult to see that behind the two last reasons lurks the image of political strife. The vindication of witches and demons here clearly means a shield against those factions that had hurled the country into the civil war, or against those, the Hobbesians, that held the foundations of government could solely be based on the power mechanizations of positive law. Against the former, Glanvill would uphold the category of disinterested reason, against the latter, his, and the Cambridge Platonists', conviction of an organic body politic (as there was an organic, not mechanical, coherence in nature). Neither can one be a reasonable being in one respect and a fanatic in another, for everything is connected with everything else. "'Truth is never alone"; all truths "hang together in a chain of mutual dependence; you cannot draw one link without many others."[11] This certainly harkens back to the old notion of coherence, yet at the same time it is a clear countermand against the disintegration of fields that was beginning to be fostered by the Royal Society or, conversely, an appeal not to let this disintegration, this specialization, happen, to look beyond, *plus ultra*, and become aware of the consequences. The time-honored idea of the chain of being looms large behind this caution, a chain that had to be broken to make possible the rise of science. At the same time, only now can we judge what it meant to have it assembled.

But to translate witches into images of political controversy is one thing, to know (or believe) that they are agents of a real threat is another. Glanvill is very firm: "The Laws and Affairs of the other World ... are vastly differing from those of our Regions, and therefore 'tis no wonder we cannot judge of their Designs ... the Devil is a name for a Body Politick, in which there are very different Orders and Degrees of Spirits, and perhaps in as much variety of place and state, as among our selves." (p. 21) The comparably simple example of the composition of light had given an intimation of the degrees and variety of things invisible; the workings of the lodestone would offer another example ("an invisible flow of atoms producing mechanically testable effects").[12] Since all truths "hang together in a chain of mutual dependence," it should be possible to elucidate the causes and to differentiate the workings of God from those of the devil. Glanvill's aim is to colonize the realm of the devil in order to make it as accessible as the natural world, though he knows perfectly well how risky and dangerous a venture this is. His aim is to minimize risks by a continuous progress of enlightenment. Risks and dangers are only considered real as long as a society subscribes to believe in them. Therefore it has been easy for later ages, with a changed perspective of risks, to stigmatize belief in witches as a particularly monstrous example for unenlightened minds. But it must give us pause to think how many natural philosophers, experimenters, and scientists in the seventeenth century felt the threat to be real. Perhaps only now are we in a position to judge what it means to be encircled by dangers and risks that are invisible yet produce effects that are experienced daily.

For the Cambridge Platonists, witches and demons may have been a reification of the dangers concomitant with the rise of a rigidly mechanist paradigm and with the optimism of scientific progress, for with the solution of one problem numerous others ensued. In the words of Reinhart Koselleck: "Already in the 17th century the discrepancy was discovered that exists between progress in the field of technology and civilization on the one hand and the moral attitude of man on the other. Again and again it is notified that morality does not keep pace with technology and its development."[13] To bridge this gap was clearly Glanvill's intent, be it in his vindication of witches or in his design of a new theology.

Seven years after *Plus Ultra*, in 1675, Glanvill published *Anti-fanatical Religion and Free Philosophy. In a Continuation of the New Atlantis*. In it he tried to supply what Bacon, in his fragmentary utopia, had left out: an ethic. It is, like Bacon's piece, the narrative of a traveler who recounts what the governor of Bensalem has told him. It is, however, not set in some blurred past but in the present: what happened since we first heard of New Atlantis. This gives him the opportunity to place his account of Solomon's House, i.e., the Royal Society, within the political and theological context. In a thinly disguised form we hear of he civil war which originated in zealous persons who vilified the light of reason; we get summaries of the ongoing theologico-political debates, we get arguments in favor of his own position, the Latitudinarian or Broad Church, with its principle of tolerance. The essay is, again, a vindication of the use of reason, which is seen to be compatible with religion. (The accent is on *right* reason to distinguish it from *pure* or "disinterested," "objectivist" reason; right reason is to be aided by revelation, as was also postulated by Henry More.)[14] It is, again, as much directed against the "vanity of dogmatizing" as against the various forms of prejudice; it stresses, again, the need of experiments and observations but cautions against certainties. Yet the drift of the argument sounds different. The design of the new and enlightened religion that is developed in accordance with the new science is said "to *perfect humane Nature*; To restore the empire of our *minds* over the *will*, and *affections*; To make them more temperate, and contented in reference to themselves, and more humble, meek, courteous, charitable and just towards others." (p. 25) This is to say that the new paradigm has to be accompanied, if not governed, by a new attitude of man and his mind, not in the sense of a *res cogitans* but in the sense of improving his moral nature and of making him more responsible – for himself and towards others.

This, indeed, is in striking contrast to the intention of his paragon: where Bacon plans the benefit of mankind, i.e., the improvement of man's condition on earth by technological means, Glanvill intends a reorientation of man on the strength by his reasoning power, which leads him to a soundly based morality that may even be a controlling instance over things developed elsewhere. Both thinkers are pragmatists, but whereas Bacon, the jurist and statesman, advocates the progress of science theoretically for practical purposes, Glanvill, the preacher, actually looks for its relevance in daily life. The Atlantides, the narrator tells us, had divines that "were methodical in their preaching" (the existence of preachers would have been unthinkable for Bacon!): their "preaching was

Practical ... [they] laid down the true, sober, rational, experimental method of action." (p. 44f.) This means that new thinking and new ways to act and react have to be in step; what happens in one sphere must have consequences for the other. Glanvill feels sure that with an "inlargement" of our knowledge of nature also our "uses of life," our morality, will be "greatly promoted, and advanc'd." (p. 49)

And yet he holds that nothing should be taken as certain or "establish'd." "They [the Atlantides] held no *infallible* Theory" (p. 48); with due reverence for Descartes and his "mechanical wit," yet "the *Mechanical* Principles *alone* would not salve the *Phaenomena*" (p. 51); they would have to be supplemented by the "Platonical Logoi Spermatikoi, and *Spirit* of *Nature*; and so would have the *Mechanical* Principles aided by the *Vital*." Any system-building and all it entails, however, runs counter to what Glanvill proposes. All man can do is to be on the alert, to take nothing for granted, to adapt his moral nature to the changes in science – observing, experimenting, ever aware of an impending imbalance, the risks he is taking and that are threatening him in his totality – and to act and react accordingly.

Glanvill's admonition is one of the last attempts to correct and supplement what is seen in retrospect as the new scientific paradigm by insisting on the integration of all human endeavors. He steers, as it were, a middle course between mechanistic rationalism and what may be called an illumined rationalism. His arguments are socio-religious, not socially neutral; anthropogenic, not "objective" or detached; ethical, not morally immune. Everything still *means* something and wants to be interpreted. He is, in the terminology of Foucault, still a "hermeneut," not a "functionalist."[15] From the frame of reference of what was to become the new science one may say that "various purposes or functions are conflated and confused."[16] From this angle Glanvill certainly did not contribute to the rise of science and might even have been mistaken as to what it was all about.

Yet he should not be dismissed as an extinct specimen of an outdated way of thinking. He remains a reminder that at a certain point in history something was beginning to go wrong. In his vehement criticism of Cartesian dualism, Edmund Husserl has written: "But the natural scientist is unaware of the fact that the permanent basis of the subjective workings of his mind is the world of life around him (*Lebensumwelt*). It is constantly presupposed as fundament, as working field – only here and here alone are his questions, his methods of thought, meaningful."[17] Objectivism, he contends, is naive and one-sided, for the scientist forgets that "nature" is only the product of his own mind. This is the ultimate cause of the present crisis – not only in thinking but in ethics, in politics, and elsewhere. "The rationality needed now is nothing but the really universal and really radical selfawareness (*Selbstverständigung*) of the mind in the form of an universal and responsible science in which a completely new mode of scientific procedure is set about, where all sorts of questions – questions of being and questions of norm, questions of so-called existence – find their place."[18]

To remember Glanvill, then, and bearing Husserl's admonitions in mind, means to point to problems that have been held under cover for a long time – problems, or their functional equivalents, that we can no longer afford to ignore.

Notes

1. For the various uses of "reason," "right reason," empiricism," "explanation," etc., see Lotte Mulligan, "'Reason,' 'Right Reason,' and 'Revelation' in Mid-Seventeenth-Century England," in Brian Vickers (ed.), *Occult and Scientific Mentalities in the Renaissance* (Cambridge: Cambridge University Press, 1984).
2. Richard S. Westfall, "Newton and Alchemy" in Vickers, *Occult and Scientific Mentalities*.
3. Stuart Clark, "The Scientific Status of Demonology," in Vickers, *Occult and Scientific Mentalities*, p. 356.
4. The phrase is Brian Vickers'; see Vickers, *Occult and Scientific Mentalities*, Introduction, p. 41.
5. This is the title of a book by Glanvill attacking scholastic philosophy (1661).
6. Again reissued with other essays in 1675 as "Against Modern Sadducism in the Matter of Witches and Apparitions."
7. Clark, "Scientific Status of Demonology," p. 355.
8. *De Augmentis Scientiarum*, II, 2, quoted in *ibid.*, p. 355.
9. Quoted in *ibid.*, p. 360.
10. *Ibid.*, p. 364.
11. Jackson I. Cope, *Joseph Glanvill: Anglican Apologist* (St. Louis, 1956), p. 112.
12. Mulligan, "'Reason,' 'Right Reason,' and 'Revelation,'" (p. 382).
13. Reinhart Koselleck, "Fortschritt und Niedergang," in Reinhart Koselleck (ed.), *Niedergang* (Stuttgart, 1980), p. 229.
14. Mulligan, "'Reason,' 'Right Reason,' and 'Revelation,'" p. 384.
15. Michel Foucault, *Les Mots et les Choses* (Paris, 1966), Chap. 2.
16. The quote is Brian Vickers' characterization of seventeenth-century occult philosophers as opposed to the new method based on distinctions and functional specificity. See Vickers, *Occult and Scientific Mentalities*, Introduction, p. 42.
17. Edmund Husserl, "Die Krisis des europäischen Menschentums und die Philosophie," Vienna lecture, May 1935, in Edmund Husserl, *Die Krisis der europäischen Wissenschaften und die transzendentale Phänomenologie* (1934–1937) (Den Haag: Nijhoff, 1962), p. 242f. "Aber der Naturforscher macht sich nicht klar, daß das ständige Fundament seiner doch subjektiven Denkarbeit die Lebensumwelt ist, sie ist ständig vorausgesetzt als Boden, als Arbeitsfeld, auf dem seine Fragen, seine Denkmethoden allein Sinn haben."
18. *Ibid.*, p. 246. "Die Ratio, die jetzt in Frage ist, ist nichts anderes als die wirklich universale und wirklich radikale Selbstverständigung des Geistes in Form universaler verantwortlicher

Wissenschaft, in welcher ein völlig neuer Modus von Wissenschaftlichkeit in den Gang gebracht wird, in dem alle erdenklichen Fragen, Fragen des Seins und Fragen der Norm, Fragen der sogenannten Existenz, ihre Stelle finden."

Selected Additional Bibliography

Boyle, Robert. *The Sceptical Chymist* (1661). London: Everyman Library, n.d.

Cassirer, Ernst. *Die platonische Renaissance in England und die Schule von Cambridge.* Leipzig and Berlin, 1932.

Cope, Jackson I. "'The Cupri-Cosmits': Glanvill on Latitudinarian Anti-Enthusiasm." *The Huntington Library Quarterly.*

Douglas, Mary and Wildavsky, Aaron. *Risk and Culture.* Berkeley, University of California Press, 1983.

Easley, Brian. *Witch-Hunting, Magic and the New Philosophy.* Sussex: The Harvester Press, 1980.

Glanvill, Joseph. *Plus Ultra* (1668). Reprint with an introduction by Jackson I. Cope. Gainesville, Florida, 1958.

Glanvill, Joseph. "Against Modern Sadducism in the Matter of Witches and Apparations." Essay VI in *Essays on Several Important Subjects in Philosophy and Religion* (1675). Reprint. Stuttgart and Bad Cannstatt, 1970.

Glanvill, Joseph. "Anti-fanatical Religion, and Free Philosophy: In a Continuation of the New Atlantis." Essay VII in *Essays on Several Important Subjects in Philosophy and Religion* (1675). Reprint. Stuttgart and Bad Cannstatt, 1970.

Godet, Alain. "Hexenglaube, Rationalität und Aufklärung: Joseph Glanvill und Johann Moriz Schwager." In *Deutsche Vierteljahresschrift für Literaturwissenschaft und Geistesgeschichte.* Vol. 52. Stuttgart, 1978, 581–603.

Houghton, Walter E. "The English Virtuoso in the Seventeenth Century." *Journal of the History of Ideas* 3 (January 1942), 51–73; (April 1942), 190–219.

Jobe, Thomas Harmon. "The Devil in Restoration Science: The Glanvill-Webster Witchcraft Debate." *Isis* 72 (1981), 343–356.

Newton, Isaac. *Newton's Philosophy of Nature: Selections From His Writings.* Edited by H. S. Thayer. New York, Haffner Publication Co., 1953.

Prior, Moody E. "Joseph Glanvill, Witchcraft, and Seventeenth Century Science." *Modern Philology* 30 (1932–1933), 167–193.

Reichert, Klaus. "The Two Faces of Scientific Progress; or, Institution as Utopia." In *Utopian Vision, Technological Innovation and Poetic Imagination*, edited by K. L. Berghahn and Reinhold Grimm. Heidelberg, 1990, 11–28.

Sprat, Thomas. *History of the Royal Society*, edited with critical apparatus by Jackson I. Cope and Harold Whitmore Jones. St. Louis, Washington University, 1958.

A VICTORIAN THUNDERSTORM: LIGHTNING PROTECTION AND TECHNOLOGICAL PESSIMISM IN THE NINETEENTH CENTURY

IDO YAVETZ

Tel Aviv University

In the defining framework for this volume, we read that, "toward the close of the twentieth century, technological pessimism and post-modern sensibilities appear related as elements of late counter-enlightenment cultural trends." This creates a powerful image of centuries in conflict, and depicts late twentieth-century technological pessimism as fundamentally opposed to the enlightened, progressive spirit of the nineteenth century. In this short essay, I would like to examine this image a bit more closely.

In his recent book on the British engineering profession, R. A. Buchanan observed:

> ... the only persistent theme of a belief shared by all engineers in the nineteenth century was that of an assumption of beneficent progress. This, of course, was not distinctly an engineering phenomenon, as the belief was widely shared in the period, but in so far as engineering achievements seemed to have confirmed and justified progress, engineers could be regarded as having had a particular vested interest in it, and they suffered a specially acute ideological disillusionment with the weakening of belief in progress after 1914.[1]

On the other hand, Martin J. Wiener observed in his influential book, *English Culture and the Decline of the Industrial Spirit*, that by the later nineteenth century:

> The era of the industrial revolution, which had established the order of Victorian England, was ripe for reevaluation. Disenchantment with industrialism encouraged historical reinterpretation of the origins of the machine age. Beginning in the eighteen-eighties, both scholarly and popular historical accounts of the industrial and agricultural revolutions were intertwined with the recasting of social values. This historical writing was not only hostile to unregulated capitalism, but also questioned the value of technological advance, and the pursuit of economic growth itself. Through this writing there was fixed upon the English mind a strikingly negative image of what was, in the long perspective, perhaps the most decisive contribution of England to the history of the human race.[2]

Confronted by these two characterizations, possibilities such as the following immediately come to mind. It could be that the nineteenth century idea of progress was not synonymous with scientific, technological, and industrial growth.[3] Alternatively, the views of Buchanan and Wiener are incompatible, and one of them must be wrong. Or both views could be right, but with respect to different communities; perhaps there is more to C. P. Snow's *Two Cultures* than many of us might care to admit.

History, unfortunately, does not seem to lend itself to such sharp characterizations, and different aspects of nineteenth-century history lend support to all three possibilities. In the first part of this paper, I shall examine a debate that broke out in 1888 following certain controversial statements by the British physicist Oliver Lodge about lightning and the proper protection from its effects. I shall endeavor to show how a seemingly innocent scientific dispute about lightning reveals the frustrating complexity of an attempt to understand the behavior of the British scientific and engineering communities in terms of their commitment to the ideal of progress. That is not to suggest that the lightning debate was a debate between proponents and opponents of the idea of progress in the guise of a discussion about lightning; it was a debate about lightning, argued and eventually decided on strictly scientific and technological grounds. It would be wrong to conclude, however, that various overtones of issues not directly connected with the nature of lightning were entirely banished from the public debates about Lodge's contentions. In that respect, the lightning debate may be used to extract certain general moods and views that contributed to the character of the scientific and engineering communities of the time and that were given expression in the course of the debate. In particular, I shall argue that to the extent that the debate reflected on the question of science and technology as agents of progress, it revealed conflicting notions of progress, how it should be promoted, and by whom. In the second part of the paper, I shall suggest that such ambiguities about the concept of progress were not confined to this one occasion. I shall illustrate with a few examples that the lightning debate faithfully reflected the existence of a wide range of attitudes in the nineteenth century toward the idea of science and technology as agents of progress.

* * *

In 1878 a Lightning Rod Conference was convened in London, which described its purpose as follows:

To consider the possibility of formulating the existing knowledge on the subject of the protection of property from damage by electricity, and the advisability of preparing and issuing a general code of rules for the erection of Lightning Conductors.[4]

The initiative came from the Meteorological Society, which was represented in the conference by three members, two of whom were elected president and secretary of the conference. Three more of Britain's foremost institutions joined the effort, each contributing two members: the Royal Institute of British Architects, the Society of Telegraph Engineers and Electricians, and the Physical Society. Two further "co-opted members" were elected: Prof. David E. Hughes, the famous and highly respected inventor of the carbon microphone, and Prof. W. E. Ayrton, a bright young telegraph engineer, soon to become one of England's foremost technical educators. Of the other members, two more will be important for our discussion: William Henry Preece, senior electrician to the Post Office, and George J. Symons, the secretary of the conference.

The conference was nearly four years in preparation. During this time sixty treatises on the subject of lightning and lightning protection were abstracted by the members. Many of these studies were first translated from foreign languages. Reports were obtained from eight British manufacturers who specialized in the design and installation of lightning protection systems. Finally an impressive body of case studies of apparent failure by lightning protection systems were collected. Upon examination of all this information at the end of the four years, the conference produced a report consisting of a 300-page summary of their findings and a 19-page list of guidelines to safe lightning protection. It is worth noting that the 300 pages of data summaries give the report a far less dogmatic character than the 19 pages of code might suggest. It is also interesting to note that this report is still regarded by contemporary lightning researchers as one of the most important landmarks in the history of lightning protection.[5] The theoretical underpinnings of the subject have changed dramatically in the past hundred years; current lightning protection codes are characterized by a systematic decision-making process not even remotely evident in the 1882 report. Remarkably enough, however, the basic practical guidelines put forth by the report have been incorporated into modern codes with only minimal alternations.

Nonetheless, in 1888 the very foundations of this report were dramatically challenged by one of England's rising scientific stars, Oliver J. Lodge. Lodge, a staunch supporter of Maxwell's theory, intimated that the design principles for lightning conductors then in use were fundamentally flawed. Lodge based his critique on the analysis and interpretation of a set of experiments, from which he made certain extrapolations on the nature of lightning. I shall briefly describe the setting in which he presented the sensational results of these studies, and highlight aspects of his presentation which are of importance to my subject here.

In 1888 the widow of Robert Mann funded a lecture series in her late husband's name under the auspices of the Society of Arts. The audience for the lectures seems to have been envisaged as consisting of serious students of science, who should not, at the same time, be considered specialists in the specific topics of discussion. Robert Mann had been an enthusiastic student of lightning phenomena and actively involved in the design and installation of lightning protection systems. It seemed only appropriate that the first lecture in the series should be devoted to lightning. Oliver Lodge was asked to deliver the lecture.

Peter Rowlands tells us in his book on *Oliver Lodge and the Liverpool Physical Society* that Lodge had a gift for public speaking.[6] His two Mann lectures on lightning and lightning protection undoubtedly provide a case in point. Instead of tiring his audience with prolonged theoretical analyses, Lodge presented a series of strikingly simple but highly suggestive experiments. He discharged Leyden jars across small air gaps, thus creating the electrical sparks that never failed to excite audiences ever since their discovery in the eighteenth century. In one set of experiments, Lodge made his sparks fly between two metal plates. On the bottom plate he erected little structures and showed how the sparks preferred certain paths to others in opposition to received notions, while on other occasions they showed no preference whatever when received notions made no allowance at all for such apparent arbitrariness. In another set of experiments, Lodge discharged his little sparks into conductors made of copper, the metal of choice for practical lightning conductors. He showed that rather than follow this highly conductive path, the discharges often left the conductor and jumped across several inches of air seeking another path.

The basic phenomenon that Lodge's experiments demonstrated most vividly was that electrical circuits behave very differently when subjected to a sudden rush of current than when carrying a steady current. Lodge pointed out that lightning discharges clearly involve sudden and very large currents. He proceeded to portray the accepted practices as futile because they were based on analysis of lightning protection systems in terms that were appropriate only for steady currents. One should note that most of this had already been said two years earlier by David E. Hughes, one of the original members of the 1882 conference. He concluded from a series of experiments on current pulses that during the first portion of the variable period, the current encounters an enormous obstruction. He explicitly stated that this must be the case with lightning and that lightning conductors must be designed with this in mind.[7]

Lodge, however, carried his own analysis much further than that into an attempt to reflect on the very nature of the lightning discharge and from there into a sweeping critique of all lightning protection principles. First he showed that the electrical properties of his little sparking circuits were such that each of his sparking discharges actually oscillated with frequencies of hundreds of thousands of oscillations per second. Then he concluded from an estimation of the lightning channel's electrical properties that the typical lightning flash would actually oscillate several million times per second.[8] In short, he came to the conclusion that for all practical purposes, his oscillating Leyden jar discharges were analogous to a full-fledged lightning flash.

With these general observations in hand, Lodge proceeded to reexamine the accepted view regarding the purpose of a lightning conductor. It had been customary to consider that the lightning conductor must simply divert the lightning current into ground by the shortest and least obstructive path. Lodge argued that this view can no longer be maintained. The lightning discharge cannot be regarded simply as a given quantity of electricity that must be diverted to ground, Lodge said. The lightning discharge carries a certain amount of energy, which must be dissipated before it ceases to present a danger.[9] Should the resistance of the lightning path be sufficiently high, Lodge said, the energy will dissipate in a single, unidirectional current surge. However, if the conductor is a good one, characterized by very low resistance, then "... it is not a mere one-directional rush, it is an oscillation, a surging of electricity to and fro, until all the energy is turned into heat."[10] Such surges pose the greatest source of danger, Lodge said, calling his audience's attention to Maxwell's theory of electromagnetic waves. According to this theory, rapid and violent oscillations in a struck lightning conductor would set up electromagnetic waves, which would spread out from the strike site. These electromagnetic undulations would invariably induce current surges in any metal objects nearby, and flashes between these objects, whether they were grounded or not, could never be ruled out.[11] Therefore, Lodge concluded, a modicum of resistance might actually be an advantage, and a highly conductive path such as that afforded by copper may actually compromise the safe action of a lightning conductor.[12]

It is not my purpose here to enter any further into the details of Lodge's experiment and his analysis thereof. For the present, it is only important to note that Lodge concluded his lectures with a disturbing, cautionary note. Advancing science, he said, proved the problem of lightning protection to be far more complicated than previously believed. Indeed, he suggested, practical lightning conductors, even if designed according to the new views he was advocating, would not afford perfect protection; and existing lightning protection systems should be regarded with a healthy does of suspicion.

Before examining the reaction to Lodge's lectures, it should be pointed out that the year 1888 was particularly exciting for British electrical scientists and engineers. On the scientific side, sensational news arrived from Germany that Hertz had succeeded in detecting electromagnetic waves in his laboratory. All at once, Maxwell's electromagnetic waves were no longer a paper theory; they became part of experimental reality, and, as we just saw, electromagnetic waves played an important role in Lodge's lightning studies. On the engineering side, the infamous Electric Lighting Act of 1882 was amended in 1888. Electric lighting companies could no longer be compelled to sell their assets to local government authorities after twenty-one years. The period was extended to forty-two years, and the government was now required to purchase the assets (should it elect to) at "fair market value," words which did not exist in the 1882 act. This event is undoubtedly less known than Hertz's discovery, but it is difficult to say which was deemed more important at the time. A large section of the British electrical industry regarded the 1882 act as anathema to progress. Following the amendment, *The Electrician*, Britain's most influential electrical weekly, noted with pleasure the rush of investment and accelerated development in the British electric lighting industry. Indeed, when *The Electrician* published its traditional New Year's summary of the preceding year in January 1889, most of the editorial was dedicated to the amendment of the Electric Lighting Act; not a single word was said about Hertz and electromagnetic waves.

It would probably be an unjustifiable exaggeration to suggest that Lodge's analysis of lightning protection fell on a euphoric British electrical community like a shower of icy water; yet there are indications that the somewhat pessimistic character of his conclusions may have contributed to the manner of their reception. It is, of course, only natural to expect that Lodge's radical redrawing of the basic understanding of lightning and the protection therefrom would attract attention. Just how much attention the study actually drew became dramatically apparent five months after Lodge's Mann lectures, during the Bath meeting of the British Association in September 1888. The intense debate was renewed the following January at a special meeting under the auspices of the Institution of Electrical Engineers (IEE).

During the Bath meeting, a special joint session of the mathematical and engineering sections was organized for the purpose of discussing Lodge's ideas about lightning and lightning protection. William Henry Preece, then the chief electrician of the Post Office and the president of the engineering section at Bath, spoke first and quite skilfully turned the discussion into a war of words.[13] Preece opened by praising the virtues of the 1882 Lightning Rod Conference report. He then likened Lodge to the Biblical character Balaam the soothsayer, who was commissioned to curse the army of Joshua and ended up blessing it instead. Like a reverse image of his Biblical predecessor, who rode his ass to Joshua's encampment and reneged on his contract, so did Lodge ride the British AAS (Association for the Advancement of Science), cursing the Lightning Rod Conference of 1882 instead of blessing it as he was expected to do. Preece then proceeded to state that he could not comment with authority on Lodge's mathematics and on the theoretical analysis he offered for his experiments. He did inform his audience, however, that he had over 500,000 lightning rods under his direct supervision and that this vast source of practical experience gave him full confidence in the efficiency of existing lightning protection systems.

The only reasonable conclusion from this, Preece suggested, is that for some reason, which he did not pretend to understand, Lodge's oscillating Leyden jar discharges were not analogous to the natural phenomenon of lightning. No practical conclusions could therefore be drawn from them, and they could not be put forth as justifying the sort of criticism Lodge levelled at the design principles underlying practical lightning protection systems. Mathematics employed judiciously in the service of practice could be a useful tool, Preece continued. But the fanciful theoretician who allows mathematics to take control is not merely useless, but a source of dangerous misconceptions. Practical electricity owes nothing to mathematicians, Preece concluded. Practice, not mathematics, is the only guide to practical engineering.

Lodge, in his turn, quickly proved himself more than a match for Preece in public debate. Lodge first pointed out that he was never commissioned to bless or curse the 1882 Lightning Rod Conference, and hence did not think any breach of contract occurred. Furthermore, he said, there were three characters in the biblical story Preece conjured: Balaam, Balak, and the ass. Balak could be the president of the Society of Arts, who asked Lodge to deliver a lecture; Balaam, he continued without elaboration, was the prophet who said what he ought to have said; now he was not sure who the ass was, unless it was the party who spoke against the prophet. Lodge did acknowledge the problem of justifying the analogy between laboratory discharges and lightning flashes but maintained all the conclusions he drew from it, and just so there would be no mistaking his meaning, he intensified his attack on the 1882 conference.

In sharp contrast with Lodge's highly charged response, Kelvin simply said that he decided not to take any of Preece's remarks on mathematics seriously and that he was not angry with Preece because Preece never really made anyone angry anyway. Then, ever so gently, he put forth his opinion that, as it stood, Lodge's analysis was incomplete. Although it raised many interesting and disturbing questions, he could not draw any definite conclusions from it. Five months later, when Lodge's work was reexamined at the IEE, Kelvin reiterated the same opinion. Referring to the 1882 report, he stated,

... I do think that this book continues practically to hold the field, by its practical rules and recommendations for the rendering of buildings and telegraphic apparatus safe against lightning.[14]

When George J. Symons' turn came in Bath, he added a new dimension to the debate. The former secretary of the 1882 Lightning Rod Conference began by acknowledging his incompetence to make detailed judgements on scientific matters. He stressed that he had no intention whatever to question Lodge's experimental ability. In fact, *The Electrician* reported that "... he yielded to no one in appreciation of Prof. Lodge as an experimenter." However, he pleaded with the audience to keep in mind that Lodge's

... were, after all, simply laboratory experiments, and it seemed to him that what they wanted was something on an infinitely larger scale. ... He certainly did feel most desirous ... that they should not simply, on the strength ... of those laboratory experiments, allow it to go forth from this great Association that there was any uncertainty in the protection of the public buildings throughout the country (applause). It seemed to him that a very serious responsibility attached to those who would make a suggestion of that nature.[15]

To Symons, then, Lodge appeared an alarmist. This seems to have struck a chord, and when the discussion was resumed at the IEE, Preece gave a much more direct expression to this view. The language of the Lightning Rod Conference summary was aimed at ordinary people, he said, to instill in them a sense of security in the face of lightning; so, said Preece sarcastically,

What does Professor Oliver Lodge do? He labels every lightning protector with a great notice "Beware of this; it bites, it fizzes, it spits, it does all sorts and kinds of dangerous things; do not trust it, it is of no use. Sleep in your house in comfort, and do not mind these Lightning Rod Conference people."[16]

What further aggravated this image is the fact that neither in his Mann lectures nor at Bath did Lodge offer any practical amendment to the 1882 code, which he criticized with uncompromising fervor. When the discussion was resumed at the Institution of Electrical Engineers, under the watchful eye of its new president, Kelvin, Lodge did offer a tentative set of alternative practical design principles. Instead of alleviating the problem, however, this attempt by Lodge to comply with practical demands only added more fuel to the fire. The differences between Lodge's principles and the nineteen-page code formulated in 1882 turned out to be only minor. His critics were quick to focus on the glaring discrepancy between the rebellious nature of the conclusion that Lodge drew from his experiments on one hand, and the conformist character of his practical design guidelines on the other. Lodge's critics could now maintain even more ardently that his experiments were largely irrelevant to the practical business of improved lightning protection, and they used that as a further indication that Lodge's laboratory work somehow failed to capture the realities of nature unbound.

It would be impossible in the course of this short outline to go over each of the responses in this protracted debate. Several engineers took the floor; some reiterated their refusal to consider the Leyden discharges as dependable miniature lightnings and repeated Preece's observations on the general effectiveness of existing systems; others recalled various instances of unexplainable failures. Two other reactions, however, are worthy of special notice.

American representation in this meeting came in the authoritative form of Henry Augustus Rowland, America's foremost Maxwellian physicist. *The Electrician's* report shows that Rowland did not mince words and stated crisply that Lodge's work inspired him with distrust. "The question seemed to be whether that experiment actually represented the case of the lightning. He [Rowland] was very much disposed to think that it did not."[17] Rowland also found fault with Lodge's theoretical analysis of sparking under rapid oscillations, not only with the attempt to extrapolate its lessons to the case of lightning.

The final word in the Bath debate was left to the president of the mathematical section, Lodge's close friend and one of Maxwell's most illustrious followers, George Francis FitzGerald. Significantly enough, his reaction was quite similar to Rowland's. FitzGerald joined Preece, Rowland, Hughes, Symons, and many others in doubting that the experiments caught the underlying reality of lightning, and said he was disposed to believe that in many ways the experiment "... was not at all analogous to a lightning flash." In a short one-to-one with Lodge, FitzGerald was forced to admit that he had not read Lodge's most recent paper on the subject of lightning, but this did not save Lodge from having to acknowledge that the experiments he demonstrated indeed failed to answer some of the issues FitzGerald raised. FitzGerald closed the discussion with unabashed support for the 1882 report.

... he did not think there was any doubt at all, quite independently of any experiments of Prof. Lodge's, ... that as a matter of experience lightning conductors had protected buildings, whatever the explanation of it was, and that there was no doubt that they had been right on the whole [in their methods of erecting lightning conductors.][18] Perhaps there were improvements possible; perhaps they might be able to protect themselves from those unfortunate discharges that occurred in telephones and so forth; but as a whole lightning conductors had been a great protection to mankind from danger from lightning (applause). That was undoubtedly the result of their experiences.[19]

In conclusion of this necessarily short glimpse at the lightning debate, let me first call attention to the fact that Lodge's lightning research was severely criticized by both practical engineers and theoretical scientists – especially theorists with a strong commitment to Maxwell's electromagnetic field view. On first examination it may seem as though this harmony of opinion reflected, besides a critique of some theoretical and practical aspects of Lodge's work, a commitment on the part of both scientists and engineers to a fundamental, shared value, which Lodge somehow attacked and which others now sought to defend. In the next section I shall examine the extent to which the lightning debate is suggestive of a more or less coherent concept of progress, which Lodge may have undermined and which his critics may have wished to uphold.

* * *

Lodge's scathing attack on lightning protection practices seems to have come at a particularly inconvenient time for British electrical engineers. Throughout the 1880s British engineers in general were a favorite scapegoat for all those concerned with the increasing technological gap between Britain and the United States. The electrical engineers in particular were vulnerable to this sort of criticism because the electrical industry, more than any other, symbolized technological progress. The main focus throughout the 1880s switched from telegraphy to electric lighting – an area in which British performance seemed particularly poor. In 1888, however, the Electric Lighting Act was amended, thus removing what appeared to be a major obstacle to expanding electric lighting throughout England. There was new reason for hope, and British electrical engineers were even harder pressed to prove their ability by quickly furnishing Britain with electric lights. An unexpected attack on their competence from another direction may have been highly irritating in its implication that British engineering was always, somehow or other, failing as an agent of true progress.

At the same time, Lodge also intimated that advancing science revealed the problem of lightning protection to be far more intractable than previously thought. He based this observation on a set of experiments which he interpreted, in part, directly on the basis of Maxwell's electromagnetic field theory. In their response, the engineers seemed to have effectively criticized the ability of Lodge's theory-assisted laboratory science to deal effectively with nature. This, in turn, could have been undesirable to other scientists, especially Maxwellian theorists. After many years during which Maxwell's theory enjoyed the dubious honor of being regarded merely as a paper theory, Hertz provided direct experimental evidence in its favor. Under these circumstances, the apparent discrepancy between Lodge's criticism and the lessons of experience may have been precisely the last thing Maxwellians like FitzGerald and Rowland wanted. To them, Lodge may have been dangerously undermining the sensational success of Hertz. At this moment of triumph for pure science, Lodge's lightning investigations could too easily be regarded as casting doubts on the very ability of pure science to deal usefully with the realities of nature. From this viewpoint, individuals like FitzGerald and Rowland may be seen as facing an unpleasant choice: they could support their colleague Lodge, but then undermine science's claim to being a co-sponsor of progress; or they could maintain this claim at the price of debunking, *on scientific grounds*, Lodge's lightning analysis. It appears that under the circumstance, they elected the second choice. In short, to both the scientific side and the engineering side, Lodge may have appeared as an obstructor of progress. Accordingly, the negative reaction he elicited on both sides may reflect a shared concern that scientific and engineering knowledge should not betray the cause of progress.

It would be tempting, perhaps, to leave things at that; but, unfortunately, the details of the debates at Bath and the IEE make it difficult to uphold this view without qualifications. To begin with, one should avoid placing excessive emphasis on the relevance of a general nineteenth-century notion of progress to the lightning debate. Considering the vehement openness with which the Victorian scientists and engineers gave expression to their ideas in the course of the debates, the fact that the word "progress" was not mentioned throughout the lengthy discussion should be taken as a caution. Furthermore, I already noted that several engineers did support Lodge's critique with recollections of failures by lightning protection systems. On the other hand, these observations should not serve as an excuse for dismissing any role whatever for concepts of progress to the lightning debate. Rather, we must consider more carefully how such notions could fit in given the above observations. Many recent studies in the history of technology suggest that there is good reason to regard engineering and science as characterized by different states of mind, though there seems to be little agreement on what precisely defines the differences.[20] The evolutionary model of engineering knowledge outlined by Walter G. Vincenti in his recent book, *What Engineers Know and How They Know It*, suggests that such distinctions may actually vary in time.[21] Elsewhere I have shown that the lightning debate was only one of several controversies that reflected a restructuring of the relations between the scientific and engineering communities.[22] At the core of these changing relations stood the emergence of engineering science, marked by research aimed at formulating basic principles for the development and maintenance of progressively more sophisticated technological systems. Throughout this process, apparent debates between theoreticians and "practical men" actually reveal a deeper commitment by both engineers and scientists to the idea of "the harmony of theory and practice," to use the words of W. J. M. Rankine's.[23] On the other hand, both sides were just as strongly committed to maintaining their own sense of independent identity during a period of rapid change. Thus, while engineers and scientists of the period showed considerable agreement on the required harmony of theory and practice, they also displayed deep tensions on other levels. This is strikingly revealed by an angry editorial in the mechanical engineering journal, *The Engineer*, which took the lightning debate in Bath as a good opportunity for berating the scientists.

... we should be lacking in our duty did we not assert whenever the occasion for assertion arises, that the engineer, and the engineer alone, is the great civiliser. The man of science follows in his train.[24]

The dividing issue, then, seems to be the identity of the real agent of progress. It seems highly unlikely that devoted scientists like FitzGerald and Rowland, who never belittled the importance of practical experience gained by engineers, would meekly accept this harsh verdict. At the same time, they did not, indeed could not, take credit for many of the developments effected by engineers. More likely they would point out that the engineer's idea of progress may not be the same as the scientist's. After all, scientific advancement generally means more comprehensive understanding of natural processes. To the engineer, advancement fundamentally means a greater ability to harness nature's forces for the purpose of changing man's living conditions. This point was well expressed by J. A. Fleming in his memoirs.

In the Charter of the Institution of Civil Engineers of London, the aim of Engineering is defined as that of controlling the great energies of Nature for the use and benefit of mankind. Nevertheless it has been well said that Nature is a Coy mistress, and will not be wooed simply for the sake of her dowry of applications. Scientific research must therefore go forward independently of any possible practical applications, and it is certain that when thus pursued in a disinterested and reverent spirit it is one of the very highest occupations of the human mind.[25]

It was, indeed, H. A. Rowland who rebuked members of the American Association for the Advancement of Science for their narrow-minded preoccupation with productivity at the expense of promoting the cause of progress in pure science.[26] From this point of view, then, the apparent unison between the scientific and engineering critics of Lodge's lightning work seems to be motivated less by a *shared* notion of progress, and more by each party's desire to preserve and assert its own *independent* interpretation of the word. As Hugh Aitken expressed it:

Technology, to oversimplify, is always for hire; pure science if it is true to its own ideas, never is. The difference is not one of morality. The plain fact of the matter is that the rules of the game which have been found historically to be best suited for progress in science are not identical to the rules of the game best suited for the advance of technology.[27]

It seems then, that we have gone full circle. The lightning debate landed us right back between the two observations by Buchanan and Wiener, highlighting the question that they raise. On one hand, certain aspects of the lightning debate suggest a wide commitment to a rather vague idea of progress; on the other hand, it appears that scientists and engineers did not share a single coherent view regarding the meaning of progress and how best to serve it. The editorial in *The Engineer* provides one glaring case in point, and very likely Preece and Rowland would have interpreted the word "progress' in different, and in this case incompatible ways. From this point of view it will be seen that Lodge was criticized by scientists and engineers not because he challenged one widely shared concept of progress, but because he happened to challenge at one and the same time the otherwise divergent notions of progress held by scientists and engineers. In this manner, the lightning debate serves to illustrate the possibility that "progress" in the nineteenth century may have signified very different, sometimes conflicting things to different people. Now, if nineteenth-century scientists and engineers did not share the same ideas of progress, could not similar or even sharper disharmonies exist between other communities? Furthermore, if we looked closer at either the engineering or scientific communities, would we not find further subdivisions within their respective ranks?

Martin J. Wiener amassed an impressive body of evidence showing that within certain literary and political circles the idea of technology as the agent of progress was received with suspicion and often with open hostility. There is no need to repeat here Wiener's evidence regarding the skeptical attitudes of J. S. Mill toward productivity and economic growth,[28] or Ruskin's striking nightmare vision of the twentieth century.[29] In the remaining space, I would like to suggest with the help of a few illustrative examples that the situation was not merely one characterized by simple arguments for or against technology or science. In fact, we find a complex range in which different, often competing, notions of progress appear, with some people advocating certain technologies as opposed to others. All this raises the possibility that no coherent agreement existed in the nineteenth century on what the word "progress" really meant.

* * *

Already in 1829 we find early explicit recognition of the possibility that technological progress may have its dark side. In *Signs of the Times*, Thomas Carlyle noted what he saw as a disturbing tendency on the part of his contemporaries to organize their lives according to mechanistic images. Mechanization did not stop with the sailors who replaced muscle and wind power with steam power, he wrote. He observed that some of his contemporaries seemed to regard education as a mass-produced commodity, to be manufactured according to the principle of the production line without proper regard to varying individual needs. With bitter sarcasm he pointed out how his contemporaries were trying to boost their artistically impoverished culture by the creation of public institutions for the manufacture of future artists, as if to nourish the feeble spirit of art by mass production. One cannot help noting the similarities between the merciless, sleepless and impersonal policing agency of "public opinion" in Carlyle's "age of mechanism," and the equally impersonal personification of a mechanical society's will in the form of the ceaselessly watchful eye of Big Brother in Orwell's 1984. With such images Carlyle cautioned his contemporaries to beware lest their mechanized society itself become one big inhumane machine.

Half a century later, but across the channel in France, one finds a pessimistic depiction of the industrial future of the 1950s in Albert Robida's (1848–1926) widely read *The Twentieth Century* (1883) and *The Electric Future* (1892). In satirical prose bolstered by magnificent drawings from his own pen, Robida depicted the world of the 1950s as a veritable electrical hell. Brittany, widely regarded as the most backward province of late nineteenth-century France, is depicted by Robida as the last bastion of hope in the electric beehive reality of the future. Without machines and without electricity, backward Brittany turns into the coveted resort to which the inhabitants of the 1950s flock on their vacations; it is described as a paradise of green pastures and hills – an island of humanity in a world gone insane with technological progress.

Not everyone, however, was so sweepingly critical. Consider again W. E. Ayrton, who served on the 1882 Lightning Rod Conference. He could hardly be portrayed as a technological pessimist. He was an ardent advocate of technological growth and improvement. It would, however, also be wrong to suggest that he considered all forms of technological growth equally welcome. In fact, Ayrton was a devoted campaigner for electric power but seems to have considered steam power as something almost evil whose use should be minimized. Ayrton consistently pushed for the use of hydroelectric energy in order to contain and reduce the widespread use of coal and steam. When he addressed an auditorium filled with working men during the 1879 meeting of the British Association in Sheffield, he expressed these ideas in colorful, romantic language.

We are, it is true, a commercial people, but do we not still love our hills and our fields? . . . why should not the mountain air that has given you workmen of Hallamshire in past times your sinew, your independence of character, blow over your grindstone now? Why should not division of labour be carried to its end and power brought to you instead of you to the power? Let us hope, then, that in the next century electricity may undo whatever harm steam during the last century may have done, and that the future workman of Sheffield will, instead of breathing the necessarily impure air of crowded factories, find himself again on the hillside, but with electric energy laid on at his command.[30]

Furthermore, while Ayrton clearly favored hydroelectric power over steam and coal, he was judiciously cautious even here. In the 1890s he advised against the plan to build a hydroelectric plant over the Victoria Falls on the Zambezi river. He strongly criticized the idea of destroying the natural beauty of the falls for the sake of an overambitious engineering project. Would Ayrton have addressed the working men of Sheffield as he did were there not already a significant degree of discontent among them about being robbed of their preindustrialized living conditions? Ayrton, one should note, is reputed to have been a highly principled man; would he have castigated the foul air of crowded factories had he not himself believed that this consequence of industrialization was a turn for the worse? Was his concern for preserving the natural beauty of the Victoria Falls a posture deceitfully adopted for the sake of mere appearances, or a cynical ploy aimed at advertising electric power? Possibly, but to my mind not very likely.

With all these qualifications, Ayrton still considered electricity to be the salvation of the future, but one need not go outside the electrical press to find evidence that electrical progress was not deemed progressive by all. Ever since the electric light made its debut, a fierce contest was fought between the incipient electric lighting industry and the older and well-established gas industry. The various strategies used in this unrelenting advertising campaign were not limited to counting the ability or inability of the competing industries to provide better lighting. Safety, pollution, and notions of general aesthetics played an equally important part. Thus, when progress came to Paddington Station in the form of electric arc lights, the following rendition of a famous nursery rhyme appeared in Britain's foremost electrical weekly, *The Electrician*, penned by an anonymous gas enthusiast:

Twinkle, twinkle little arc,
Sickly, blue uncertain spark;
Up above my head you swing,
Ugly, strange expensive thing!

Now the flare gas has gone
From the realms of Paddington,
You must show your quivering light,
Twinkle, blinkle, left and right!

When across the foggy air
Streams the lightning's purple glare,
Does the traveler in the dark
Bless your radiance, little Arc?

When you face with modest blush,
Scarce more bright than farthing rush,
Would he know which way to go
If you always twinkle so?

Cold, unlovely, blinding star,
I've no notion what you are,
How your wondrous "system" works,
Who controls its jumps and jerks.

Yours a lustre, like the day!
Ghastly, green, inconstant ray!
No; where'er they worship you,
All the world is black or blue.

Though your light perchance surpass
Homely oil, or vulgar as,
Still (I close with this remark)
I detest you, little Arc![31]

Scientific developments during the second half of the century also seem to undermine rather than enhance the image of the nineteenth century as guided by adherence to a more or less coherent idea of progress. Among scientists, as among engineers, we can find various notions of scientific progress, some of them rather bitterly opposed to one another. J. A. Fleming, the electrical engineer who invented the diode and who admired Maxwell, observed in 1934:

In looking back over that period of time, the last half of the nineteenth century and the first third of the twentieth, over the greater part of which my memories of scientific work extend, one thing is very noteworthy in comparing the present with the past, and that is the enormous increase in our practical technical achievement, and on the other hand the diminished confidence we now have in the validity of our theoretical explanations of natural phenomena. There is, in short, at present far less "cocksureness" in scientific thought that we have plumbed the depths of knowledge or reached the end of possibilities in theory and practice.[32]

Fleming's observations leave the impression that while diminished confidence in scientific knowledge was the order of the day during his later years, there had been an earlier period of "cocksureness." In order to further examine this image, we may turn to some comments made in 1910 by Oliver Heaviside, the brilliant interpreter of Maxwell's theory who made major contributions to the theory of telegraphic and telephonic transmission lines. The first part of Heaviside's statement reveals one of the reasons for Fleming's shaken confidence, namely, the rise of special relativity as the proper kinematical basis for Maxwell's theory. The second part of the statement, however, indicates that whereas Einstein's contribution did not invoke an optimistic response from Heaviside, Heaviside's own contributions to Maxwell's theory did not instill a sense of growing optimism in all nineteenth-century scientists.

Some do not believe in the materiality of the ether. This view is thoroughly anti-Newtonian, anti-Faradaic, and anti-Maxwellian. What mean action and reaction, the storage of energy, the transit of force and energy through space &c., &c., if there is no medium in space? For space is nothing at all, save extension. Lord Kelvin used to call me a nihilist. That was a great mistake, ... He was most intensely mechanical, and could not accept any ether unless he could make a model of it. Without the model he could not accept electrodynamics to be dynamical. But I regard electrodynamics as being founded upon Newton's dynamics of the visible, and resting upon it throughout.[33]

What Kelvin regarded as nihilistic about Maxwell's theory was strikingly illustrated by Maxwell himself in 1879, when the youthful J. A. Fleming was attending his classes.

In an ordinary belfry, each bell has one rope which comes down through a hole in the floor to the bellringer's room. But suppose that each rope, instead of acting on one bell, contributes to the motion of many pieces of machinery, and that the motion of each piece is determined not by the motion of each rope alone but by that of several, and suppose, further, that all this machinery is silent and utterly unknown to the men at the ropes, who can only see as far as the holes in the floor above them. Supposing all this, what is the scientific duty of the men below?[34]

By measuring the forces and motions of each rope, Maxwell explained, the operators below could arrive at a complete dynamical description of this intricate bell system, but:

This is all that the men at the ropes can ever known. If the machinery above has more degrees of freedom than there are ropes, the coordinates which express these degrees of freedom must be ignored. There is no help for it.[35]

In other words, Maxwell was suggesting that scientists may have to learn to live with the distinct possibility that a fully specified mechanical model of the ether – Maxwell's invisible machinery – may forever resist scientific scrutiny; perhaps they should not even consider it their duty as scientists to provide such an explanation. Kelvin's bitter opposition to such notions is well documented.[36] Already in 1884 he poignantly expressed his opinion that to the extent that Maxwell's theory was associated with the ascendancy of such views, it was a step backward.[37] Consequently, when in 1889 Kelvin hailed Hertz's discovery of electromagnetic waves as a great scientific achievement, he already harbored grave doubts about certain characteristics of Maxwell's theory. For our purposes here it is important to note that Kelvin's reaction was not isolated. Similar reactions throughout the scientific world of the late nineteenth century were persistent enough for the historian J. T. Merz to record the feeling of "something lost" with the adoption of Maxwell's electromagnetic field theory.[38] As Oliver Heaviside noted in 1910, the fact that Maxwell's electromagnetic field theory seemed to replace explicit mechanical processes with the principles of generalized dynamics found little favor with Kelvin. To Heaviside this must have seemed an obstruction to scientific progress, but Kelvin may have sensed that Maxwell's theory was only the first step toward even deeper doubts and more sweeping revisions of physical knowledge. Indeed, we have seen that as Heaviside was asserting his position against Kelvin in 1910, he was already fighting against developments that the very success of Maxwell's theory brought in its wake and that Heaviside certainly would not have categorized as progress. As we go backward in time from Fleming's 1934 assessment, to Heaviside's rejection of the emerging new basis for electromagnetism in 1910, to Kelvin's deep-seated doubts in 1884 about the nature of electrodynamics represented by Heaviside's work, it becomes progressively more difficult to portray late nineteenth-century science as guided by adherence to some global notion of progress. Rather, as new scientific ideas came to the forefront, they seemed to entail revisions of the idea of progress in science which were not equally palatable to all scientists. Developments in the 1880s that Heaviside deemed progressive brought forth a pessimistic reaction from Kelvin; by 1910 Einstein's contributions turned Heaviside's optimistic anticipation of a Maxwellian future into a pessimistic feeling that science had taken the wrong track.

France of the 1880s supplies one more indication that among biologists as among physicists, what some regarded as progress, others saw as depressing turns for the worse. This is how Jean Henri Fabre, the illustrious French entomologist, contrasted his behavioral studies with what some of his biologist colleagues were doing.

You rip up the animal and I study it alive; you turn it into a object of horror and pity, whereas I cause it to be loved; you labour in a torture-chamber and dissecting room, I make my observations under the blue sky to the song of the Cicadas; you subject cell and protoplasm to chemical tests, I study instinct in its loftiest manifestations; you pry into death, I pry into life. And why should I not complete my thought: the boars have muddied the clear stream; natural history, youth's glorious study, has, by dint of cellular improvements, become a hateful and repulsive thing.[39]

It is well worth mentioning that Fabre's image of insect studies afield was enthusiastically embraced by the very same generation of entomologists who otherwise severely criticized his relentless attack on the theory of evolution.[40]

Finally, returning to attitudes toward technology, it is worthwhile to point out a different, perhaps more disturbing form of pessimism, poignantly expressed by the penultimate experience H. G. Wells arranged in 1895 for his time traveler. Traveling far into the future, the time traveler encounters the Eloi and the Morlocks – the decadent evolutionary descendents of the late nineteenth-century aristocracy (the Eloi) and working class (the Morlocks). At first sight the Eloi appear to be pure, innocent, and frail creatures without any technological capabilities, seemingly living freely off unmolested nature, wholly vegetarian and incapable of any violence whatever. At first sight the Morlocks appear to be ugly, murderous, machine-using underground dwellers. Then comes the horrible realization that there is nothing natural about the Eloi's manner of subsistence, that in effect the repulsive but industrious Morlocks are artificially maintaining the Eloi like "fatted cattle" and actually feeding upon their flesh. In 1829 Thomas Carlyle could still hope that the technology of the age of mechanism could be contained within a limited sphere of the human experience, H. G. Wells in the electrified 1890s already saw technology and social life as inseparably linked. Rejection of technology, Wells seems to imply, will not lessen our dependence on it; it will only drive technology underground, rendering our relationship with it and with each other parasitic rather than symbiotic.[41]

* * *

In conclusion, there can be little doubt that as R. A. Buchanan noted, there was wide commitment in the nineteenth century to the idea of progress. Looking at the lightning debate, however, we found it difficult to ascribe a single coherent notion of progress to the scientists and engineers who took part in it. Further evidence suggested that the ascendancy of any particular technology or scientific idea elicited optimistic reactions from some individuals and deeply pessimistic ones from others. Which aspects of nineteenth-century enlightened notions of progress are we currently rebelling against, then? Are our rebellious *fundamentally* different from those of the previous century? Do our current sensibilities represent a novel, counter-progressivist and antitechnological reaction to all that is past, or are we continuing a well-established tradition of assimilation by criticism of new technologies into our ever changing social and cultural context? Are we all simply technological pessimists, or are we perhaps, like W. E. Ayrton, pessimistic about certain technologies and optimistic about others? It would seem that these questions require a satisfactory answer before we can regard late twentieth-century postmodern technological pessimism as heralding a new alternative to enlightened nineteenth century progressivism.[42] "Progress," without elaboration, still continues to be a most useful word in political campaigns throughout the industrialized western democracies of our day; and it appears that the question of whose "progress" the word refers to still continues to be just as difficult to answer now as it was one hundred years ago.

Notes

1. R. A. Buchanan, *The Engineers: A History of the Engineering Profession in Britain, 1750–1914* (London: Jessica Kingsly Publishers, 1989), p. 188.
2. Martin J. Wiener, *English Culture and the Decline of the Industrial Spirit, 1850–1980* (Cambridge: Cambridge University Press, 1981), p. 82.
3. J. B. Bury, in the introduction to *The Idea of Progress: An Inquiry into its Origin and Growth* (London: Macmillan and Co., 1920) sees "progress" as standing first and foremost for liberty, democracy, and the secular belief in this world as opposed to the religious belief in the next. Industry, technology and science are important only in so far as they serve this more fundamental cause. Only two of Bury's nineteen chapters are devoted explicitly to developments in science and technology: the chapter on Bacon and the idea of increasing knowledge, and the chapter on the Great Exhibition of 1851. Industrialization and technological growth, Bury suggests, are important primarily for having given progress a concrete existence in the mind of the average man (p. 324). At the same time, he notes H. de Feron's observation in 1867 on the occasion of the Paris Exhibition that all the spectacular inventions of the day were equally capable of serving evil and good causes (p. 331). Thus, respectable voices already exclaimed that technological advance is not synonymous with progress.
4. *Report of the Lightning Rod Conference,* edited by the Secretary, G. J. Symons, F.R.S. (London: E. & F. N. Spon, 1882), p. viii.
5. R. H. Golde, *Lightning Protection* (London: Edward Arnold, 1973), p. 1.
6. Peter Rowlands, *Oliver Lodge and the Liverpool Physical Society* (Liverpool: Liverpool University Press, 1990).
7. *The Electrician* 16 (February 5, 1886), 256.

8. The main difference between a real coaxial line and the lightning path in Lodge's analysis was that instead of the metallic return in the form of a conducting sheath, Lodge assumed the return current to be a displacement current in the dielectric surrounding the flash. He assumed (without a reasonable explanation) that the area of cloud discharged by a typical flash has a diameter of about fifty meters (current estimates are at least an order of magnitude larger). Thus, the return displacement current would flow in a cylindrical volume whose external diameter would be fifty meters. Taking the flash channel to be a very good conductor, Lodge further concluded that because of the impulsive nature of the lightning discharge, the current therein is conducted on the surface, according to the principle of skin conduction experimentally corroborated by Hughes in 1886. Thus Lodge considered that the inside of the channel would remain largely unmagnetized. From this he calculated the inductance of the path assuming that the magnetic induction associated with the channel and return currents would be confined to the area between the inner and outer radii. (Oliver Lodge, *Lightning Conductors and Lightning Guards* (London: Whittaker & Co., 1892), pp. 87–97).

9. "This fact is often forgotten by lightning-rod men; they speak as if there were a certain quantity of electricity to be conveyed to earth and there was an end of it; but they forget the energy of the electric charge, which must be got rid of somehow." (*The Electrician* 16 (February 5, 1886), 39.)

10. *Ibid.*, p. 38. Lodge repeated this point many times, constantly emphasizing this notion of the dissipating role played by the lightning conductor. "A noteworthy though obvious thing is, that the energy of the discharge must be got rid of somehow. The question is, how best to distribute it. It is no use trying to hocus-pocus the energy out of existence by saying you will conduct the charge to earth quite easily and quietly. Conducting the charge to earth is no secure mode of getting rid of the energy, and unless the energy is exhausted the charge will rise up again, and so may swing up and down a good many times before the store of energy is all gone; and nothing can be worse as regards disruptive effect than this repeated and violent passage of an enormous electric current." (*Ibid.*, p. 13). In 1890 he still said just as emphatically, "The problem of protection . . . ceases to be an easy one, and violent flashes are to be dreaded, no matter how good the conducting path open to them. In fact, the very ease of the conducting path, by prolonging the period of dissipation of energy, tends to assist the violence of the dangerous oscillations." (*Ibid.*, pp. 368–369). All this is possible only under the tacit assumption that the resistance of the lightning conductor is comparable to that of the air channel from cloud to ground. At least on one occasion, Lodge's work suggests that he actually believed this to be the case. Based on his transmission line model of the lightning channel, Lodge proved that a channel resistance of 15000 Ohms is not high enough to prevent oscillations. He then argued that this value far exceeds any reasonable guess at the real resistance, since the resistance of the lightning conductor must not exceed a few Ohms lest it become comparable to that of the air path. (*Ibid.*, p. 93). It would seem from that comment that Lodge may have thought the air channel is characterized by a resistance of no more than a few hundred Ohms.

11. *Ibid.*, pp. 402–403.

12. Thus Lodge writes: "A thin iron wire is nearly as good as a thick copper rod; and its extra resistance has actually an advantage in this respect, that it dissipates some of the energy, and tends to damp out the vibrations sooner. Owing to this cause a side flash from a thin iron wire is actually less likely to occur than from a stout copper rod." (*Ibid.*, p. 371).

13. For a nearly verbatim account of the Bath discussions, see "Discussion on Lightning Conductors," *The Electrician* 28 (September 21, 1888), 646.

14. *Journal of the Institution of Electrical Engineers* 18 (1889), 520.

15. *The Electrician* 21 (September 28, 1888), 677–678.

16. *Journal of the Institution of Electrical Engineers* 18 (1889), 459.

17. *The Electrician* 21 (September 28, 1888), 676.

18. The sentence in brackets appears as an integral part of FitzGerald's comments in the BA report of the 1888 meeting. See *Report of the British Association for the Advancement of Science, 1888*, p. 615.

19. *The Electrician* 21 (September 28, 1888), 680.

20. It is not possible to record here the extensive literature on the relationship between science and technology. For a useful survey of this question, see J. M. Staudenmaier, *Technology's Storytellers: Reweaving the Human Fabric* (Cambridge, Mass.: MIT Press, 1985), pp. 83–120.

21. Walter G. Vincenti, *What Engineers Know and How They Know It* (Baltimore: The Johns Hopkins University Press, 1990), p. 245.

22. I. Yavetz, "Oliver Heaviside and the Significance of the British Electrical Debate," *Annals of Science* (in press).

23. See David F. Channell, "The Harmony of Theory and Practice: The Engineering Science of W. J. M. Rankine," *Technology and Culture* 23 (1982), 39–52.

24. Oliver Lodge, "Sketch of the Electrical Papers in Section A at the Recent Bath Meeting of the British Association," *The Electrician* 21 (September 21, 1888), 622.

25. J. A. Fleming, *Memoirs of a Scientific Life* (London: Marshall Morgan & Scott Ltd., 1934).

26. Daniel J. Kevles, "Rowland, Henry Augustus," *DSB* 13.

27. H. G. J. Aitken, *Syntony and Spark: The Origins of Radio* (New York: John Wiley & Sons, 1976), p. 18.

28. M. J. Wiener, *English Culture and the Decline of the Industrial Spirit, 1850–1980*, pp. 32–33, 90–92.

29. *Ibid.*, p. 27.

30. W. E. Ayrton, "Electricity as a Motive Power," *The Electrician* 3 (September 20, 1879), 215.

31. *The Electrician* 21 (October 19, 1888); 771 (quoted from *St. James Gazette*).

32. Fleming, *Memoirs of a Scientific Life*, p. 242.

33. Oliver Heaviside, *Electromagnetic Theory* (New York: Chelsea Publications Inc., 1971), Vol. 3, p. 479.

34. J. C. Maxwell, *Scientific Papers*, 2 Vols., ed. W. D. Niven (Cambridge: Cambridge University Press, 1890), Vol. 2, pp. 783–784.

35. *Ibid.*

36. S. P. Thompson, *Life of William Thomson, Baron Kelvin of Largs*, 2 Vols. (London: MacMillan and Co., Limited, 1910), Vol. 2, pp. 1012–1085; C. Smith and M. N. Wise, *Energy and Empire: A Biographical Study of Lord Kelvin* (Cambridge: Cambridge University Press, 1989), pp. 488–494; M. N. Wise and C. Smith, "The Practical Imperative: Kelvin Challenges the Maxwellians," in R. Kargon and P. Achinstein (eds.), *Kelvin's Baltimore Lectures and Modern Theoretical Physics: Historical and Philosophical Perspectives* (Cambridge, Mass.: MIT Press, 1987), pp. 323–348.

37. "... it seems to me that it is rather a backward step from an absolutely definite mechanical motion that is put before us by Fresnel and his followers to take up the so-called electromagnetic theory of light in the way it has been taken up by several writers of late." (Kelvin's first Baltimore Lecture, in R. Kargon and P. Achinstein (eds.), *Kelvin's Baltimore Lectures and Modern Theoretical Physics*, p. 12.)

38. John Theodore Merz, *A History of European Thought in the Nineteenth Century*, 4 Vols. (New York: Dover Publications, Inc., 1965 – a republication of the original 1904–1912 publication), Vol. 2, pp. 93–94.

39. Jean Henri Fabre, *Life of the Fly* (New York: Dodd, Mead and Company, 1913), pp. 4–5.

40. For an analysis of Fabre's attitude toward the theory of evolution, see I. Yavetz, "Jean Henri

Fabre and Evolution: Indifference or Blind Hatred," *History and Philosophy of the Life Sciences* 10 (1988), 3–36; and *idem*, "Theory and Reality in the Work of Jean Henri Fabre," *History and Philosophy of the Life Sciences* 13 (1991), 33–72.

41. H. G. Wells, *The Time Machine* (London: Heinemann Educational Books, 1895), pp. 74, 80.

42. Bill Luckin's depiction of the "anti-pylon" movement in the late 1920s and early 1930s, illustrates antitechnological environmental language which bears resemblance to environmentalist campaigns in the 1980s. At the same time, it also shows that there was a great range of opinions within both the preservationist and electrical progressivist camps of the 1920s, with varying degrees of willingness to arrive at compromises (Billy Luckin, *Questions of Power* (Manchester: Manchester University Press, 1990), pp. 95–98.) Compare, for example, Luckin's account of pro- and antitechnological arguments in the 1920s with the typical exchange between Thomas R. de Gregori, "The Back to Nature Movement: Alternate Technologies and the Inversion of Reality," *Technology and Culture* 23 (1982), 214–217, and John Bryant, "The Back-to-Nature Movement, Modern Technology, and the Inversion of Dr. De Gregori," *Technology and Culture* 23 (1982), 218–222. For a quick sketch of the complexity and range of current opinions, showing in particular that there still appears to be a great deal of energy behind highly optimistic points of view, see Andrew Jamieson, "Technology's Theorists: Conceptions of Innovation in Relation to Science and Technology Policy," *Technology and Culture* 30 (1989), 505–533, especially 531–533. For another brief discussion of technological optimism and pessimism from the nineteenth century to the 1980s, see John G. Gunnell, "The Technocratic Image and the Theory of Technocracy," *Technology and Culture* 23 (1982), 392–416.

WHEN BAD THINGS HAPPEN TO GOOD TECHNOLOGIES: THREE PHASES IN THE DIFFUSION AND PERCEPTION OF AMERICAN TELEGRAPHY

MENAHEM BLONDHEIM

Hebrew University of Jerusalem

The telegraph was supposed to bring about a world of good. When Samuel Finley Breese Morse tried to persuade the American government to promote his telegraph, he argued that "the greater the speed with which intelligence can be transmitted from point to point, the greater is the benefit derived to the whole community." Few nineteenth-century Americans would have quarreled with his rationale. When Morse, forecasting the ultimate impact of his invention, invoked the image of making "one neighborhood of the whole country," contemporaries might have suspected "the crazy painter" of exaggerating the blessings of his invention, not of pessimistically anticipating the discontents of a future mass society.[1] Perceptions and evaluations of the telegraph changed radically in three distinct phases that paralleled the transformations of telegraphic technology in the process of its diffusion.

I

Early on, telegraphy appeared as an unqualified blessing to American society. Not only did the ultimate impact of the telegraph, as predicted by its promoters, appear benevolent; the technology on which it was based and its nature as a medium added to its benign aura.[2]

Indeed, the sharp and overwhelmingly positive preliminary image of the telegraph owed something to the Lord Almighty. Some characteristics of the first practical application of electrical phenomena to public use were explicable to nineteenth-century Christians only in language they were accustomed to hear in their churches.[3] Both God and, to an extent, the electric current were omnipresent: they could be at different places at the same time. They both were invisible and inscrutable yet their effects were far reaching. Besides, no one appeared to know all that much about either. God's ways were notoriously vague and his deeds inexplicable, nor could most Americans understand how the telegraph actually worked or even what that "mysterious agency" – electricity – really was.[4]

Not only did the technology employed in telegraphy send good vibrations, its basic function appeared more elevated and possessed more éclat than other inventions of the time. Other inventions improved the process of making or moving things. The telegraph, however, performed no function as mundane as shooting bullets, separating cotton seeds from fibers, or transporting goods and passengers; it was expected to facilitate the "extension of knowledge, civilization and truth." Telegraphy was, after all, "the application of electricity, with lightning speed, to the transmission of thought," and thought, in turn, represented "the highest faculty of man made in the image of God." Hence, the telegraph catered to "the highest and dearest interests of the human race."[5]

Yet the positive reception of the telegraph was qualified: telegraphy elicited wonder but no real excitement. Early promoters, trying to pay their way by public performances of the telegraph and telegraphy, found meager audiences. Nor did the promotional practice of charging visitors to telegraph offices a half-cent to have their names transmitted to the other end of the wire, and a whole cent to have it sent back, yield many pennies. Even discounting possible scepticism, the telegraph in operation was a disappointment; "there was no interest or excitement about the marvelous instrument," lamented an eyewitness to early American telegraphy.[6] Unlike the mechanical novelties of the time, nothing in the performance of the device helped make the scientific and technological principles it was based on intelligible, nor was the sight of the apparatus at work enthralling.[7]

More generally, there was something remote about the telegraph. The device itself stood oracle-like in the exclusive custody of its operators, the operating room out-of-bounds to customers. No familiarity nor rapport was possible between machine and those whom it served, as was the experience of later Americans with their home telephones. No emotional involvement, such as some in our day have with their capricious PCs, could develop with telegraph machines. Locomotives could be nicknamed, with individuality and familiarity, *Tom Thumb* or *Old Ironside*. Invoking the image of awesome, magnificent, but distant "Lightning" was as familiar as contemporaries would go in tagging the telegraph.

Furthermore, the telegraph's properties as a medium, as well as its nature as a device, promoted detachment. A frequently told story of a blushing maiden entering and leaving the telegraph office with the same sealed, perfumed love letter in hand illustrated one aspect of the telegraph as a cold and impersonal medium. The brevity of telegraphic messages (because transmission costs

increased with each word), the exposure of messages to operators and func-
tionaries at both ends (and even along the wire route), and the total elimination
of the personal features of the writer's hand accentuated the impersonal, remote
aura of telegraphy, making it a subject of awe rather than of affection.

Above all, the telegraph's transcendence of local boundaries enhanced the
air of foreignness and aloofness that became identified with the device. Even
though it may appear paradoxical, the mere act of communicating with distant
points instantaneously appears to have accentuated the notion of distance rather
than that of proximity. Numerous descriptions of early use of the telegraph
record the communicators commenting on the great distances that separated
them, at the same time they were marveling at the novel rapid connection;
they spoke of communicating with "friends who seem so near but are indeed
so far away". The unity of communication and transportation, from prehistory
to telegraphy, probably helped make instantaneous meetings of minds across
space accentuate the physical distance of bodies.[8]

However unfriendly, the early telegraph was still deemed a useful extension
of man in one of his most social activities – communication. Contemporaries
shared Morse's reasoning that "whatever facilitates intercourse between the
different portions of the human family will have the effect ... to promote the
best interests of man." Even Henry David Thoreau, at his crankiest, could find
nothing worse with the telegraph than its power to occupy the American mind
with news of Princess Adelaide catching the whooping cough, or, more generally, to demonstrate that distant parts of the country may have nothing important
to communicate to each other. The consensus still held that Morse's device
represented "the great invention of the age," and most contemporaries expected
it would "work wonders."[9]

It was not very long, however, before some disturbing voices were heard in
the discourse concerning the telegraph, which originally had been an exuberant
version of what Leo Marx calls "the rhetoric of the technological sublime."
Many Americans were apparently finding that the "most wonderful climax of
American inventive genius" proved better in the anticipation.[10]

II

"The sooner the [telegraph] posts are taken down the better," proclaimed the
Charleston Courier a mere two years after the first American line proved
successful. This opinion apparently was shared widely. "The law may make
it a penitentiary offence to break down the wires," acknowledge the *Courier*,
"but in the present state of public opinion, no jury could be found to convict

any one of the offence." The *New Orleans Commercial Times* claimed that it merely reiterated "the sentiments of nineteen-twentieths" of a community whose patience was fast wearing out, when it expressed the "most fervent wish that the telegraph may never approach us any nearer than it is at present."[11]

Such calls for eschewing the telegraph, notwithstanding its scientific merits, were raised in response to some very real evils that the telegraph had unexpectedly wrought. A celebrated incident, occurring soon after the first line emanating from Washington had reached Jersey City, directed attention to the potential prostitution of the invention. The Whigs were holding their nominating convention in Philadelphia, and the New York press rushed into print one morning with news of Zachary Taylor's nomination. As it turned out, the notice was premature; the voting did not take place until the following day. An investigation revealed that the newspapers' authority had been a dispatch by a ring of lottery swindlers, transmitting a set of numbers from Philadelphia to their accomplices in unsuspecting New York. The message was accidentally received by the press, mistakenly deciphered, and printed.[12]

That the press should have been led to publish erroneous information on authority of what some ultimately dubbed the "tell-lie-graph" was hardly a novelty. The New York papers had earlier dutifully mourned the passing of Massachusetts Senator John Davis. The senator, alive and well in Washington, had "after some days' absence reappeared in his seat," a contingency that according to the press correspondents' cipher book was to be reported by the four letters: d-e-a-d. The dispatch was read literally by journalists in New York, Philadelphia, and Boston, rather than deciphered. While Davis reportedly was delighted with the public evaluations of his career, others found such incidents of real concern. They feared the specter of intentional manipulation of public information, especially for speculation in financial and commodity markets.[13]

Indeed, verifiable evidence of such misuse of the telegraph soon became abundant. Speculators reaped "the advantage of time" in perpetrating their schemes by sins of commission – sending misleading quotations to the public – and of omission – cutting telegraph lines after early reports of prices were received by their partners. Then too, telegraph enthusiasts who were nimble enough to grasp its usefulness in apprehending fugitive criminals or summoning help to prevent crimes, soon had to concede that it could likewise help criminals perpetrate their crimes and spread false alarms to mislead their pursuers.[14]

Such demonstrations of the detrimental effects of telegraph diffusion posed serious problems for thoughtful Americans. Contemporaries were conditioned to expect that breakthroughs in science and technology would both represent and effect progress. Since the telegraph was generally considered the "mighty engine which will revolutionize the world" it practically followed that it was "destined to be a great instrument for good." The problem was compounded by the godly aura the revolutionary technology had assumed. Thus, the indignant cry of "sacrilege" was the response that the *Buffalo Commercial Advertiser* had to offer, properly exercised by the success of an ingenious stratagem by speculators to rig the corn market. The speculators, lamented the *Advertiser*, had "convert[ed] into a vehicle of fraudulent speculation an agency that seems but little less than divine." As difficult as it was to associate evil with this tamed application of divine lightning, with the heavenly technology itself, contemporaries realized clearly that "there are evils somewhere."[15]

The obvious solution that presented itself struck at the core of the belief in progress. More and more minds had to concede that telegraphy was perhaps a "tremendous engine," but also a neutral one. It was merely a means that was "capable of doing the greatest service to mankind" or "the greatest harm"; it could serve good ends or bad ones to the same great effect. Here, then, was a significant challenge to the popular notion that technological progress and human progress marched in unison.[16]

Just when American opinion was reevaluating the merits of the telegraph and, to an extent, reconsidering the established notions of the relationship of technological progress to social good, significant changes in the state of American telegraphy were under way. These changes in the nature of telegraphy highlighted dimensions of the technology not previously comprehended; and they made for a reevaluation of its function and uses. They also suggested a new locus for locating blame for those "evils" that were "somewhere."

III

Early perceptions and evaluations of the telegraph and telegraphy reflected the performance of the first American lines. These lines connected pairs of more or less distant points and made possible continuous two-way communications between them. The diffusion of telegraphy in America took off slowly, but after two years or so the industry began expanding in earnest. At first, and ever since, observers failed to realize how large scale diffusion altered the performance and even modified the nature of telegraph technology.[17]

The first stage of telegraphic transformation in the process of its large-scale diffusion was hardly conspicuous. It implied the addition of way stations between terminal points. As the pioneer line emanating from Washington gradually pushed northwards, Philadelphia, then Norristown, Doylestown and Summerville, and finally Newark and Jersey City, as well as intermediate points, were allowed to communicate with each other instantaneously. The same process accompanied the establishment of the Buffalo–New York line and other developing routes. As new stations were opened and potential users in the vast hinterlands joined the "electric link," the traffic of messages along the lines increased exponentially. Way stations and terminal points now vied with each other for occupation of the trunk line. The operators, formerly busy coding and decoding at either end of the wire, now had to perform a new and significant task – they had to manage the traffic of messages along the line. Were that not enough, the search for order on the main lines soon assumed a new dimension.

In the summer of 1846 the isolated lines pushing north from Washington, south from Albany and Buffalo, west from Boston, and east from Harrisburg and later Pittsburgh, converged on New York. The lightning lines were transformed into a network, and New York emerged as its "one grand center."[18] Managing the traffic of messages through the maze of routes and stations that had emerged called for more comprehensive solutions than those provided by existing arrangements. The problem of network management was clearly beyond the operators on the several lines; it required higher sources of authority. Telegraph leaders now had to add to their role as entrepreneurs and builders the role of system managers. They had to implement schedules and rosters that would expedite the flow of messages through the network. Clearly, they had to set up a system that would determine who would occupy the wires and when.[19]

Once telegraphers began responding to this challenge, two important changes in telegraphy took place. First, the mode of communicating by telegraph was transformed; the two-way, synchronous application of telegraphy had to be abandoned. With strict rosters and timetables for individual stations to take their turn at the wires and dispatch their piles of accumulated messages, the interactive "conversation over the wires" was no longer practicable. The "conversation" mode of telegraph usage gave way to the one-way asynchronous "message" mode, and the last vestige of intimacy was stripped from telegraphic communication. Friends could no longer occupy distant offices at the same time and exchange sentiments, and parties could no longer negotiate in real time through the great two-way "highway of thought." Instead, the telegraph was used mainly to transmit news reports, market information, and orders to buy or sell. To a far lesser extent, it was also employed to announce arrivals and delays, marriages and deaths. The one-way order of message came to characterize communication by telegraph, no longer much of the "social hormone" Marshall McLuhan had so expansively extolled.[20]

More important, through the process of networking, telegraph management emerged as an agency for controlling, not just facilitating, communication flow. Like the canal and turnpike, the telegraph constituted a conduit; like the steamship or stage coach, the telegraph filled the function of a vehicle; and like the post office and express companies, it scheduled and managed the traffic of messages. The telegraph thus became a combination of the hardware for transmitting intelligible signals and messages, and the software responsible for managing the flow of message throughput.

Scrutinizing the new state of telegraphy, contemporaries soon realized that their discontents had little to do with the technology per se, but a lot to do with the new modes in which it was operated. The hardware was fine, it was the software that was responsible for the evils. As that very small minority of Americans who used the telegraph well knew, the communicator merely handed his message to a clerk, and responsibility for all the mysterious process that followed, including the coding, transmission, decoding, and delivery, rested exclusively with the company. Most notably, it was the company that was responsible for the order in which messages were transmitted. The telegraph, unlike the mailbag, could transmit only one copy of a single message at a time. In telegraphic communication, in which celerity was the key to usefulness, the crucial decision on the order of transmitting competing messages was in the hands of company officers. Put simply, company policy and managers' discretion determined whether the messages of the evil speculator or of the honest merchant would have priority (and it goes without saying that one man's speculator is another man's honest trader).

Telegraph managers, human and fallible, conspicuous, and indeed central to the operation of the telegraph system, became the lightning rod for dissatisfaction with the operation of the lightning lines. They were found to hold too much power, over too many sensitive interests, and to wield it with a high hand. "When the enterprise was in embryo," the *Utica Daily Gazette* reminded its readers, "it was objected to on account of the facilities it would give speculators." Now, contended the newspaper, it was the directors who were objectionable, for by their methods of ordering messages they "leave nine-tenths of the community to be preyed upon by the remainder." These directors, who clearly were not "actuated ... by a wish to promote the public good" stood forewarned by the *Gazette*: "We really hope the directors will take this matter into consideration." Pointing out means of retribution by politicians, press, and the public, the newspaper informed the directors that in fact "it is manifestly their interest to protect the public." "Let the directors be warned in time," concluded the *Gazette*.[21]

In just as belligerent a mood, the *New York Sun* warned the managers of the New York and Boston line, widely suspected at the time of telegraphic malpractice, that "we do ask fair play and we will have it. The abuses of this great vehicle of public intelligence must and shall be corrected or the companies entrusted with its management shall cease to exist."[22] Similarly, the *New York Herald*, after enumerating instances in which public confidence in telegraph management had been shaken, warned "the directors of all companies" that in the future they

will be obliged to use every expedient that can be devised, to prevent not only breaches of the wire but also breaches of confidence, of the latter of which we have recently had some instances . . . When such an important method of communication as this is in the hands of those who sacrifice public benefit for private interest, it becomes a dangerous matter.[23]

On another occasion, the newspaper put the case even more bluntly: "The conductors or proprietors of these telegraph lines," snarled the *Herald*, "no doubt imagine that they are beyond the reach of the people or the power of the press, but they may find out their mistake when it is too late.[24]

All precedent, as well as the realities of the telegraph industry, supported contemporaries' identification of management with ownership. It was not surprising, therefore, that the issue of telegraph ownership shortly came under close scrutiny. What was found was hardly reassuring. The several early lines had been established and managed by separate companies, all pampered by generous acts of incorporation.[25] Even though the companies were nominally separate, the Morse patentees received a controlling interest in all of them in exchange for permission to use the patent rights. The concessions in the acts of incorporation and the exclusivity of Morse's patent in effect represented a legally conferred monopoly. And monopoly had acquired a rather nasty reputation in America, owing to the excesses of market revolution and the politics of Andrew Jackson. The evils of monopoly as exercised by telegraph promoters, owners, and managers presented contemporaries with a sufficient and necessary explanation for the mounting frustrations and discontent wrought by diffusion of the telegraph.

Amid the heaps of abuse, the barrage of complaints, accusations and denunciations directed at telegraph potentates, cooler minds during the late 1840s and 1850s attempted to sort out the real problems of control over telegraphy and to reach reasonable remedies. Whig commentators favored government control over telegraphy as the ultimate solution. Those who feared the concentration of power in government more than in the hands of private sector monopolists, called for stringent control by the legislature, the press, and the public over the management of the telegraph. The simplest solution, however, appeared to be competition in telegraphy.[26]

This last proposition was easily and understandably the most popular. Yet as subsequent developments were to demonstrate, it was based on a misperception of the real nature of telegraphic networks and of telegraphic monopoly. Competing telegraph inventions, such as Bain's electro-chemical system (precursor of the fax) and House's printing telegraph (precursor of the telex) were available in the late 1840s. Public pressure on the courts ensured that they would not be ruled as infringements of Morse's patent. Yet what many contemporaries did not realize, and what ultimately prevented the development of successful competition, was the nature of the telegraph networks as natural monopolies. To those who failed to grasp this aspect of telegraphy, developments in the industry in the course of the 1850s demonstrated an irresistible movement towards oligopoly. By the time the Civil War broke out, the country was dominated by three powerful regional monopolies.

IV

The end of the Civil War marked a more perfect union of the country and of its telegraph system. By 1866 the Western Union Telegraph Company had managed to absorb the nation's other regional telegraph networks and emerged as the nation's first private sector industrial monopoly. This landmark in industrial conquest represented much more than change in market position or corporate structure. It signified a new phase in the practices and perceptions of American telegraphy.

One important aspect of the emergence of the telegraph as a national monopoly has indeed received the attention it deserves. By allowing an efficient traffic of communications to flow throughout the Union, telegraphy could facilitate the integration and smooth management of large operations. The transportation sector was the first to grasp this potential. A well-coordinated national communication network made it possible to control efficiently and manage vast, complex flows of freight and passengers, as demonstrated in the management of railroad systems of continental scope. Then, together with the great transportation systems it facilitated, telegraphy provided the infrastructure that supported large-scale, national enterprises in production and distribution. Indeed, as Alfred Chandler has suggested, the telegraph became an indispensable tool in the hands of professional managers to coordinate and control American industries and markets. More recently, James Beniger, from an even broader perspective, assigned to the smooth running of a nationally integrated telegraph system a key role in what he considered the most significant development of the times – the control revolution that swept America into the information age. Thus, the consolidation of telegraphy as a fully integrated national institution both prefigured and facilitated the transformation of America into a truly modern national society.[27]

The emergence of Western Union as a national monopoly was also the culmination of a process of change in the nature of telegraphic communication itself. Developing gradually in parallel to the movement towards consolidation and "bigness" in the industry, a new mode of telegraphic communication to supplement the preliminary "conversation" mode and the subsequent "message" mode became highly effective. This new mode constituted, as a newspaper account of the 1850s described it, the sending of messages "broadcast over the state."[28] Technically, the broadcast mode was simple and had been in use since the inception of telegraph networks; it was performed by uniting a number of stations into a single circuit, so that a message sent by any one was registered by all. Through such an arrangement, a single source could access a broad field with uniform, simultaneous information.

Western Union's performance as a national system and its structure as a national monopoly added a new dimension to this practice of "broadcasting" messages. Western Union had reached its monopolistic position gradually, through a sequence of mergers and buy-outs of competitors. In the process, it had accumulated enormous reserves of redundant lines so that its transmission capacity far exceeded the country's needs. The company likewise retained its monopoly by buying out one potential competitor after another before they reached the crucial stage of establishing national networks. Thus, the increase in surplus transmission capacity outran the increase in demand for telegraphic services. The company was therefore capable of providing comprehensive circuits for those who wished to hire them and broadcast to the country at large.[28]

Two contemporary institutions found this potential for national telegraphic "broadcasting" particularly useful. One was the Associated Press – the news wire service that came to dominate national news-gathering and news flow by the mid 1850s. Structured as a coalition of regional telegraphic news associations, led by the New York City Associated Press, the AP became the nation's first private sector national monopoly by the late 1850s. The way the wire service operated was simple enough. Over the maze of telegraphic routes crisscrossing the nation, it gathered into its New York headquarters all noteworthy news. After processing this information, it prepared a consolidated news report representing a synopsis of major news stories. This news report was then broadcasted to member newspapers throughout the country, in one writing, over comprehensive telegraphic circuits that Western Union had connected for the purpose.[30]

Hardly known, but no less important an enterprise, the Gold and Stock Telegraph Company also realized the telegraph's potential to broadcast to the nation. Gold and Stock was originally a New York City telegraph company, which supplied local merchants with quotations from the major financial and commodity exchanges by means of a network of telegraphic indicators and printers. In the late 1860s Western Union turned on the small company, and by 1870 it controlled a majority of its stock. The giant then merged the subsidiary with its Commercial News Bureau, Western Union's controversial market information operation. Gold and Stock, in charge of what became Western Union's Commercial News Department (CND), specialized in transmitting spot news from American and European markets, instantaneously to subscribers all over the Union. Not only was CND strategically located in the major commodity, produce, and financial exchanges of the country and well positioned vis-a-vis the Atlantic cable, its breaking news, from any quarter, was given precedence over all other traffic on Western Union wires. Once the central office gave the appropriate signal, the whole network of American telegraphs was united for a brief time into a national CND circuit, and the report reached each and every telegraph station serving CND subscribers.[31]

CND and the Associated Press realized the full potential of a national telegraph network to access the entire American public with information. The one could clearly affect public opinion in general, while the other specialized in sensitive economic information. Their power for abuse was tremendous, and it appears that in some instances it was actually exercised. They represented the ultimate specter of emerging new centers accumulating enormous power in a mass national society.

In Congress, reform-minded representatives and senators, alarmed by what appeared to be real threats to American institutions, repeatedly took issue with the Associated Press and the CND during the 1870s and 1880s.[32] "Burdensome, oppressive, and dangerous," one concerned representative noted, the management of disseminating commercial information made the telegraph "the controlling agency of commerce." He found that there was "no industry, no interchange of commodity, no value, that is not at its mercy."[33] The power of the national news monopoly, according to other legislators, could have a similarly potent and just as sinister effect on American politics; it could "fatally pollute the very foundation of public opinion."[34] This was a contingency of an unprecedented "degree of importance," one committee reported, since in America "the perpetuation of the government must have its ultimate guarantee in the intelligence of the people."[35] Indeed, reported a House Committee in 1875, the telegraph monopoly had become "complete, controlling, . . . and dangerous to the public welfare."[36]

Even with all this sentiment, neither legislature nor executive effected any reform of American telegraphy. The repeated expressions of alarm in government were not enough to muster public sentiment or support to back action. The general public did not appear overly concerned with these new threatening bases of power founded on telegraphic technology. Like the process of telegraphic transmission, the operation of these forces was invisible, their power was based on the "mysterious agency" of network and system. The Western Union, the Associated Press, and the CND were nonlocal and remote, and they gave power to interests that were not well recognized in a modernizing society. Only the keenest and shrewdest observers of the late nineteenth century realized the extent of the power that became attached to institutions and leaders in the field of communication and information processing, and understood how that power was allocated to parties who were advantageously positioned in the system of social arrangements.[37]

V

Early enthusiastic appraisers of American telegraphy may be divided into those who emphasized continuity and evolution, and those who perceived revolution, a complete break with the past. To the former, the telegraph represented a sequel to the harnessing of steam for transportation, to new canals and better locks, to improved stage coaches and better pikes. It was a great and significant invention, but one that "illustrated the times," a crowning achievement of the age of invention. The latter were impressed by the discontinuities, the new departure that the telegraph represented. They considered the human harnessing of divine lightning, the first practical application of electricity, the separation of transportation and communication, to have heralded a new era.[38]

Indeed, both these perspectives had validity. In some ways the major stages of telegraph diffusion only complemented other social and economic developments in nineteenth-century America. These stages – the evolution of the telegraph line, the creation of telegraphic network, and finally, the emergence of a national system of telegraphs – represented the accomplishments of the inventor, the entrepreneur, and the corporation. These personae, so characteristic of the Early Republic, the Age of Jackson, and of the Gilded Age respectively, managed in turn to secure their respective positions by institutions crystalized in each of those periods: patent rights, acts of incorporation, and natural monopoly.

Ultimately more significant was the way in which each of the three phases of telegraph diffusion represented a break with the past and an exposition of novel forces that were to transform America. Accordingly, pessimistic observers grasped the great forces and the extreme dangers that each mode of telegraphy had unleashed. They understood that in its preliminary "conversation" mode, telegraphy exposed the locally confined to distant and sinister powers, allowing them to prey on the community. Telegraphy thus represented the forces that were working to transcend local boundaries and ultimately break down the bonds of community. Some grasped that the telegraph's "message" mode was congruent to the emergence of a market economy and national society, and finally, there were some who understood that by making "broadcast" of uniform information to the nation possible, an important foundation of mass society had been laid.

It was left to later observers to brood over the implications of the extension of telegraphic technology into the international arena, the possible fulfillment of Morse's original vision of the telegraph shaping a global village. Then some recent observers, with the advantage of perspective gained from a century and a half of understanding electric media, might have imagined the groundswell of reaction. They may have conceived the significance of the shift in the evolution of communication technology from broadcasting to "narrowcasting," by means of the casette, CATV, Demand TV, LANs, etc. The most imaginative, with optimism, pessimism or neither, may have anticipated communication technologies working to break down mass societies along class lines, ethnic identities, or cultural diversities, just as people the world over were coming to share one global information environment.

Notes

1. U.S. Congress, House, *Telegraphs for the United States*, H. Doc. 15, 25th Congr., 2nd sess., 1837, p. 30; U.S. Congress, House, *Electro-Magnetic Telegraphs*, 25th Cong., 2nd sess., H. Rept. 753, 1838, Appendix C.
2. This observation is based on a survey of the files of the *New York Herald*, *New York Journal of Commerce*, *New York Sun*, *New York Express*, *Baltimore Sun*, and *Washington Globe* from May 24 to June 2, 1844.
3. Other religious aspects in perception and evaluations of telegraphy are discussed by James W. Carey, "A Cultural Approach to Communications," in *Communication As Culture: Essays on Media and Society* (New York: Routledge, 1989), pp. 13–36, as well as in other essays in that volume; and by Daniel J. Czitrom, *Media and the American Mind: From Morse to McLuhan* (Chapel Hill: University of North Carolina Press, 1982), pp. 8–14. Both were influenced by Perry Miller, *The Life of the Mind in America* (New York: Harcourt, Brace and World, 1965).
4. See e.g., *Albany Argus*, June 11, 1845; *New York Herald*, December 6, 1845. Clippings in the Henry O'Rielly Telegraph Collection, New York Historical Society, New York, NY; "The Telegraph," poem in the *Chicago Tribune*, December 20, 1847, clipping in the Morse scrap-

books, Morse papers, Library of Congress Manuscript Division, Washington D.C. Newspaper clippings in the many volumes of scrapbooks in the O'Rielly collection, together with the scrapbooks in the Morse papers, and in the Alfred Vail papers and telegraph collection, Smithsonian Institution Archives, Washington D.C., were the major source for analysis of public opinion presented below. Reference to the respective collections is made only when unidentified or undated clippings are cited.

5.   *New York Express*, quoted in the *Pittsburgh Mercantile Advertiser*, July 8, 1848.
6.   Alonzo B. Cornell, *True and Firm: Biography of Ezra Cornell* (New York, 1884), pp. 93ff.; Alfred Vail Telegraph Diary, Vail Telegraph Collection; Frederic Hudson, *Journalism in the United States* (New York, 1873), pp. 598–599. Although Czitrom, *Media and the American Mind*, pp. 6–8, argues to the contrary, and could be supported by the *Philadelphia Dollar Newspaper*, January 1, 1846, and the *Pittsburgh Gazette and Advertiser*, December 30, 1846, it appears that the crowds surrounding Morse in the first days of the first line were lured by the news of the national conventions taking place in Baltimore. More generally the gathering of crowds in telegraph offices upon their opening appear to reflect initial curiosity, which faded quickly after initiation.
7.   For a discussion of this aspect see Oscar Handlin, "Man and Magic: First Encounters with the Machine," *American Scholar* 33 (Summer 1964), 415ff.
8.   *Philadelphia Pennsylvanian*, December 30, 1846; practically all modern accounts have commented on the divorce of transportation and communication by means of the telegraph. Of these the most profound discussion is in Carey, "A Cultural Approach," and in *idem*, "Technology and Ideology: The Case of the Telegraph," *Communication as Culture*, pp. 201–230.
9.   S. F. B. Morse quoted in the *New York Tribune*, November 13, 1847; Henry David Thoreau, *Walden* (Boston: Houghton Mifflin Co., 1857), p. 36; *Pittsburgh Gazette and Advertiser*, December 30, 1846; *Philadelphia North America*, December 30, 1846.
10.   Clipping from an unidentified New Orleans Newspaper, Morse scrapbooks, Morse papers.
11.   *Charleston Daily Courier*, June 8, 1846; *New Orleans Commercial Times*, July 13, 1846.
12.   Numerous accounts of this and related episodes may be found in the O'Rielly scrapbooks, 2nd series Vols. I, II. The most reliable account is probably Alexander Jones, *Historical Sketch of the Electric Telegraph Including Its Rise and Progress in the United States* (New York: G. P. Putnam's Sons, 1852), pp. 133–134.
13.   The best account of this episode is in Jones, *Historical Sketch*, p. 133.
14.   See e.g., *New York Herald*, March 3, 1846, March 4, 1846, August 18, 1846, November 5, 1846, November 6, 1846; *New York Evening Mirror*, December 15, 1846; *Utica Daily Gazette*, November 21, 1846; *Buffalo Commercial Journal*, March 2, 1847; *Rochester Daily Democrat*, February 13, 1847.
15.   *Philadelphia North American*, December 30, 1846; *Buffalo Commercial Advertiser*, June 8, 1846; *Philadelphia United States Gazette*, May 16, 1847.
16.   *Philadelphia Inquirer*, December 6, 1846; *Buffalo Mercantile Advertiser*, July 8, 1847; *National Police Gazette*, May 30, 1846; *New York Express*, quoted in the *Philadelphia Dollar Newspaper*, June 17, 1846; "Professor Morse's Electric Telegraph," unidentified clipping, Morse scrapbooks, Morse papers: "Justice to American Genius," *New York Tribune*, undated, Morse scrapbooks, Morse papers; U.S. Cong., *Report of the Postmaster General*, Ex. Doc. 2, 29th Cong., 1st sess., 1845, p. 861.
17.   The best of the numerous accounts of the development of American telegraphy to 1866 remains Robert Luther Thompson, *Wiring a Continent* (Princeton: Princeton University Press, 1947). The fullest contemporary account is James D. Reid, *The Telegraph in America and Morse Memorial* (New York: John Polhemus, 1886); Alvin F. Harlow, *Old Wires and New Waves* (New York: D. Appleton-Century Col, 1936), although anecdotal, is also useful.

18. Unidentified clipping, Vail Telegraph Collection.
19. Marshall Lefferts, *Address to the Stockholders of the New York and New England Telegraph Co.* (New York: Snowden, 1851); *idem*, "The Electric Telegraph: Its Influence on Geographical Distribution," *Bulletin of the American Geographical and Statistical Society* 2 (January 1857), 242–264; Richard B. DuBoff, "The Telegraph and the Structure of Markets in the United States, 1845–1890," *Research in Economic History* 8 (1983), 253–277.
20. *New York Herald*, June 4, 1844, December 6, 1847; *National Police Gazette*, May 30, 1846; *Philadelphia Dollar Newspaper*, January 1, 1846; *Albany Evening Atlas*, July 6, 1846; Edward Lind Morse, *Samuel F. B. Morse, His Letters and Journals* (Boston: Houghton Mifflin, 1914), Vol. 2, p. 224; Marshall McLuhan, *Understanding Media: The Extensions of Man* (New York: McGraw-Hill, 1964), pp. 217–227.
21. *Utica Daily Gazette*, undated clipping, Henry O'Rielly collection.
22. *New York Sun*, November 25, 1846.
23. *New York Herald*, November 22, 1846.
24. *New York Herald*, November 20, 1846; cf. *New York Tribune*, quoted in Henry O'Rielly, *Caution to the Public*, undated circular, O'Rielly Telegraph Collection.
25. The corporation, originally a private entity entrusted with public duties on behalf of the community, had gradually become a means for granting public sanction to private enterprises, serving the private pecuniary interests of the incorporators. At the time the telegraph was spreading, incorporation was fast becoming a privilege free to all. Oscar Handlin and Mary Flug Handlin, *Commonwealth: A Study of the Role of Government in the American Economy* (Cambridge, Mass.: Harvard University Press, 1969), pp. 106–181; Stuart Bruchey, *Enterprise: The Dynamic Economy of a Free People* (Cambridge, Mass.: Harvard University Press, 1990), pp. 131–133, 205–210.
26. The most profound statement in support of government control was issued by the *New York Express*, June 25, 1846; other revealing statements to the same effect include *New York Herald*, October 9, 1848, January 15, 1850; *National Police Gazette*, May 30, 1846. Strong statements in opposition to government control include *New York Evening Mirror*, December 15, 1846; *New York Evening Post*, January 5, 1846; *Pittsburgh Mercantile Advertiser*, July 8, 1848; "Justice to American Genius," *New York Tribune*, undated clipping, Morse scrapbooks, Morse papers; "Opposition Telegraphs," *Buffalo Courier*, undated clipping, O'Rielly Telegraph Collection. Most common were calls for competition in telegraphy. Representative of numerous statements to that effect are the *Troy Daily Post*, September 30, 1847; *New York Tribune*, November 13, 1847.
27. Alfred D. Chandler, *The Visible Hand: The Managerial Revolution in American Business* (Cambridge, Mass.: Harvard University Press, 1977), pp. 188–200, *et passim*; James D. Beniger, *The Control Revolution: Technological and Economic Origins of the Information Society* (Cambridge, Mass.: Harvard University Press, 1986), pp. 16–25, *et passim*.
28. *Albany Argus*, undated clipping, Henry O'Rielly Telegraph Collection.
29. The best source for these developments are the Western Union Presidents' Letter Books and other corporated records deposited in the Corporate Secretary's Office, Western Union Corporation, Upper Saddle River, NJ. WU's printed annual reports for the late 1860s and 1870s provide a good summary.
30. The development of the Associated Press is described in Victor Rosewater, *History of Cooperative News Gathering in the United States* (New York: D. Appleton and Co., 1930); Menahem Blondheim, "The News Frontier: Control and Management of America's News in the Age of the Telegraph," (Ph.D. Dissertation, Harvard University, 1989); Richard A. Schwarzlose, *The Nation's Newsbrokers* (Chicago: Northwestern University Press, 1989).
31. The best sources for G&S history are its BOD's Minutes and other corporate records deposited

in the Corporate Secretary's Office, Western Union Corporation; Western Union's Presidents' Letter Books, Western Union Corporation. Short secondary accounts include DuBoff, "Telegraph and the Structure of Markets," pp. 267–269; Blondheim, "News Frontier," pp. 350–351.

32. Lester G. Lindley, "The Constitution Faces Technology: The Relationship of National Government to the Telegraph" (Ph.D. Dissertation, Rice University, 1971); Richard B. DuBoff, "The Rise of Communications Regulation: The Telegraph Industry, 1844–1880," *Journal of Communication* 34 (Summer 1984), 52–65; Blondheim, "News Frontier," pp. 455–469.

33. U.S. Congress, House, *Telegraph Lines*, H. Rept. 125, 43rd Cong., 2nd Sess., 1877, pp. 7–8. Cf. *Speech of Charles Sumner of California in the House of Representatives*, pamphlet (Washington, 1884), pp. 25–27.

34. U.S. Congress, Senate, *Connecting the Telegraph with the Postal Service*, S. Rept. 242, 42nd Cong., 3rd Sess., 1872, pp. 4–5.

35. U.S. Congress, House, *To Connect the Telegraph with the Postal Service*, H. Rept. 6, 42nd Cong., 3rd Sess., 1872.

36. U.S. Congress, House, *Telegraph Lines*, H. Rept. 125, 43rd Cong. 2nd Sess., 1875, p. 7.

37. *Ibid.*, pp. 403–417.

38. Good representatives of these diverse tendencies are: *Buffalo Commercial Advertiser*, undated clipping, Henry O'Rielly Telegraph Collection; *New York Sun*, November 3, 1847 ("the greatest revolution of modern times and indeed of all times"); *New York Herald*, May 30, 1844 ("Prof. Morse's telegraph is not only an era in the transmission of intelligence, but it has originated in the mind an entirely new class of ideas, a new species of consciousness").

ON THE NOTION OF TECHNOLOGY AS IDEOLOGY:

PROSPECTS

ROBERT B. PIPPIN

University of Chicago

Technology as a Political Problem

A central feature in the history of Western modernization has been an ever-increasing reliance on technology in manufacturing services, information processing, communication, education, health care, and public administration. This reliance was anticipated and enthusiastically embraced by the early founders of modernity, especially Bacon and Descartes, and finally (much later than they would have predicted) became a reality in the latter half of the nineteenth century. Moreover, increasing technological power proved an especially valuable asset in liberal democratic societies. The surplus wealth made possible by such power appeared to allow a more egalitarian society, even if great inequalities persisted; representatives of such technical power could exhibit, publicly demonstrate, and so justify their power in ways more compatible with democratic notions of accountability; and a growing belief in the "system" of production and distribution as itself the possible object of technical expertise seemed to make possible the promise of a great collective benefit, given proper "management," arising from the individual pursuit of self-interest promoted by market economies.

Since that time such an increasing dependence on technology has been perceived to create a number of political problems and controversies. Commentators came to see that this reliance had certain social costs, created difficult ethical problems, and began to alter the general framework within which political discussions took place. Such problems included:

i. A greater *concentration of a new sort of social power* in fewer hands. Such a concentration of power might easily become inconsistent with democratic decision making. While there are deep compatibilities between democratic values and such scientific canons as the public demonstrability of knowledge claims as well as the public benefits of the ends to which technology can be employed (e.g., health, agricultural planning, communication), it is also true that the rise of expert elites posed a certain sort of threat. With the growing sophistication of science and technology, and the difficulties encountered by a lay public in understanding evidence, demonstrations, and the ambiguities and risks inherent in the pursuit of any end, such elites grow progressively less accountable in traditional ways for the exercise of their power, shielded as they are by their claim to greater technical competence.

ii.   A simultaneous and connected *de-skilling* of the labor force through automation, and more rigid, hierarchical forms of technically efficient administration. In such cases the imperatives generated by competition can promote an increased acceptance of technically efficient monitoring techniques (e.g., typists on centralized computers, whose "backspace" or "delete" key usage is closely monitored, and who thus can be held accountable not only for what they do but also for what they could have done, or the zealous monitoring that the phone company exercises over its operators), job simplification, greater risks to worker safety in order to conform to more efficient machines, and a variety of organizational strategies, all relatively inconsistent with basic post-Enlightenment ideals of self-respect, dignity, and autonomy.

iii. A connected phenomenon, noted by such writers as Arendt and Habermas: a narrowing of acceptable topics for "public debate," thanks to a greater emphasis on policy issues as technical issues.[1] This amounts to a *"depoliticizing" of public life*, such that much political debate becomes merely a war among competing experts or an exercise in the manipulation of symbols, a wholly theatrical celebration of rival images and icons, rather than a collective and substantive deliberation about a common societal direction.

iv.   A simple increase in the *extent of administrative power* over aspects of daily life. Foucault's claims about micro- and bio-power are relevant here, as are similar claims about the kind of power made possible by data storage in medicine, government, banking, insurance, etc.[2] A recent and (to many) worrisome example has been the project to map the human genome. With this information and new diagnostic technics, it may eventually be possible for, say, insurance companies or potential employers to predict with some reliability from a simple blood test the chance that an individual will get a stroke or coronary disease, whether he smokes or drinks too much, or even, perhaps, whether he suffers too much stress, is too neurotic, or too unsociable, etc.

v.   An extraordinary new role for science and technology in *national security issues*, requiring diversion of vast resources to ever more expensive weapons research, a diversion that has seriously and perhaps permanently derailed hopes for welfare-state capitalism.

vi. Cultural complaints that the "technological tail" was beginning to wag the "human dog"; that too many areas of daily life were being modified to meet *the needs of technical efficiency*. Complaints, for example, about being reduced to a number, having to talk to answering or voice-mail machines instead of people, excessively technological and so "dehumanized" environments for birth, illness, death, or complaints about the medicalization of mental-health issues.

Technology as an Ideological Problem

Often such topics are discussed within some sort of cost-benefit framework and under the assumption of a kind of technological fatalism, i.e., that the efficiency of a technology makes continuing or ever-growing reliance on that technology inevitable, or at least unproblematically rational. Amelioration of the social costs and an exploration of options with respect to ethical dilemmas could, under such assumptions, only occur marginally as a kind of moral hope and only after the technological imperative had been basically satisfied (a situation especially obvious in a climate of worry about international competitiveness).

A more radical critique can be detected in those who understand technology itself as a kind of *ideology*. This notion is both complex and vague, and its usefulness has suffered a great deal from an increasing use of the term to mean simply a philosophy or belief. But, as a critical concept, the notion was made possible by the Kantian revolution in philosophy and its central claim that there would be "forms" or "conditions" of experience not themselves derived from experience, but "constitutive" of the very possibility of experience. It was this notion of a priori constraints on empirical experience or, more broadly, belief formation, that set the stage for Hegel's historicization of these categories, Marx's social theory and Lukacs's use of the notion of "reification" in a full-blown "ideology critique."[3]

Within this tradition, a form of consciousness or a general comprehensive categorization of experience can be ideological in any number of senses.[4] Ideological claims can be claims about the nature of reality, the significance of a social practice, the origin and legitimacy of an institution, the authority of a moral code, or many other things. To claim that any such general, fundamental orientation to the world is ideological means not only that some interconnected set of propositions about nature, others, or the cosmos is false, unsupported by evidence or argument, unproven or irrationally believed, but that such an orientation or form of consciousness somehow prevents, even renders unnoticeable, contrary evidence or argument. Consciousness itself, the way we originally take up and make sense of things, can be "false." (Of course, this resistance to criticism is not something constructed consciously and strategically, and so the question of whether there could be, or why there should be, this sort of blindness, rather than just mistaken, overly optimistic, or ethically inappropriate worldviews, is an important one for *Ideologiekritik*.)

It is controversial whether there are or ever have been "forms of consciousness" with these characteristics, but the notion, especially when applied to some sorts of religious or moral views, is *prima facie* plausible, and has also played a major role in deflationary critiques of technology (or technology rationality, or technological "promise"). Has our relation to objects been so influenced by technical instruments, the power of manipulation and production, etc., that our basic sense of the natural world has changed, and changed so fundamentally that our reflective ability to assess and challenge such a change is threatened? Has our understanding of others and of social and political life become so shaped by technical imperatives in production, consumption, social organization, daily life, and politics, that fundamental possibilities for social existence are seen only (in a narrow and distorted sense) in terms of such technical imperatives?

Thus, in general, "ideology critics" are more interested in what is *undiscussed* in the modern experience of technology, what an extensive reliance on technology, which often is presented as a value-neutral tool, itself already hides, distorts, renders impossible to discuss as an option. To see technology as an ideology is to see an extensive social reliance on technology and its extensive "mediating" influence in daily life as already embodying some sort of "false consciousness" – again, a way of looking at things not characterizable as simply a matter of false or problematic or narrow beliefs. This means that such reliance reaches a point where what ought to be understood as contingent, one option among others, open to political discussion, is instead falsely understood as necessary (i.e., the relevant options are not even noted as credible options; hence the "false consciousness"); what serves particular interests is seen, without reflection, as of universal interest; what is a contingent, historical experience is regarded as natural; what ought to be a part is experienced as the whole; and so on.

The Classical Positions and the Classical Problems

This is a large and much discussed issue, so in order to make my point I shall need to survey the terrain from a fairly high altitude. I want first to set briefly some typical sorts of claims that "technology is ideology" and some of the problems generated by such claims before introducing a general objection and then focusing on one of the most influential recent arguments, namely Habermas's.

I begin with the most complex view, one that does not use, and would deliberately avoid, all the notions of ideology critique, i.e., Heidegger's. Heidegger claims that technology embodies an "orientation to Being" and is "ideological" in the sense that, in such an orientation, Being is "forgotten." He claims that modernity itself is "consummated" or "completed" by a technological "enframing" (*Gestell*), that technology exemplifies an understanding of Being, an absolutely fundamental orientation, which completes modern subjectivism and thoughtlessness.[5] (Technology is a "worldview," or pretheoretical "horizoning" of experience, a view also roughly maintained by Ellul[6].)

Several commentators have objected to Heidegger's explanation of the "predatory" stance of the modern subject by appeal to an obscure "history of Being," in which, it appears, Being itself is responsible for its own obscuring or for our forgetting Being. Others dispute the way he dates modernity (as originating in Plato), or complain about the ambiguous practical consequences of his critique of modernity. In my view, the central problem with Heidegger's approach is that the way it addresses the basis historical questions at issue is undialectical and even a bit moralistic. For him the appeal of the modern emphasis on power, control, and the priority of the self-defining subject seems to be the result of human hubris, a self-assertion that often sounds more like a theological account of the Fall than an historical explanation. The possibility that the basic ontological dimension embodied in a technological worldview (the subject-object split) could have been *provoked* historically or was required in some sense, given the unavoidable and genuine deficiencies and dead ends created by the premodern tradition (understood in its own rather than later terms), is not considered by him.

To be sure, he has his own, infinitely complicated reasons for this neglect, having to do with his own understanding of the History of Being and how such a history, and his (Heidegger's) own role in the destruction of Western metaphysics, play roles in the origin of modernity. But I simply note that his own position requires a historical narrative that has been, rightly I think, the subject of much criticism. (I have attempted a fuller assessment of Heidegger's position elsewhere.)[7]

I should also note that Heidegger's approach also ought to remind us of very speculative claims much discussed recently: that the modern fixation on technological power is not uniquely modern or a distortion of anything, but some sort of culmination of the deep connection between *all* knowledge and "the will to power" or a final revelation of the nature of power/knowledge (*pouvoir/savoir*).[8] These Nietzschean and Foucaultian themes, while continuous with Heidegger's approach, in some ways go much farther than Heidegger, who always seems to want to preserve a contrast between the modern "age of the world picture" and some possible alternative. These approaches also raise their own famous questions about what sort of critique or critical knowledge is possible under such assumptions, but that would introduce in this context a major digression.

To return to positions more customarily identified within the "ideology critique" tradition, we should recall the original Marxist attack: the capitalist claim for its own "rationality," understood as technical efficiency, is ideological.

First, Marx focuses on the organization of production under liberal or so-called "free market" capitalism. What are asserted to be the imperatives of technical efficiency are "functionally" ideological in Geuss's sense. The claim for the efficiency of the capitalist organization of technology is only temporarily true. When maintained beyond the early phase of capitalism, such claims become a "socially necessary illusion," functioning to mask social contradictions, actually to impede the development of the forces of production and even greater technical efficiency, and to stabilize and sustain forces of domination that could not be sustained without such illusions.[9]

This position has been most often criticized for its historical limitation to the liberal phase of capitalism. With the onset of state intervention in and management of the economy, ideology critique could no longer be a critique of political economy alone. (There was no such thing any more as "the economy" operating under its own laws. The model of a "base" supporting and, by virtue of its autonomous development, straining against and then being constrained by superstructure has been eclipsed with the arrival of welfare or state-interventionist capitalism.)

Second, the idea that the growth of productive forces is inherently emancipatory, or helps expose the ideological character of the justification for historically outmoded relations of production, has been rendered obsolete. The main contribution to growth in the forces of production now comes *from* science and technology, designed and implemented by managers and bureaucrats. Adherence to the imperatives of technical efficiency now helps legitimate the entire self-regulating social system.

Third, this notion of system now seems more relevant to social analysis than traditional class-conflict notions. Genuine, clear-cut oppositions between class interests are now rarer. Allegiances are secured through a complex and efficient system of rewards and leisure time, and all perceive themselves to be helping to operate a system of benefit to all, rather than serving the interests of a discrete, identifiable group. This means, for someone like Habermas, that reflection can identify an emancipatory interest distorted or repressed in such a system only as a *species interest*, as an interest of humanity, denied or regulated by such systematic imperatives. (Again, he wants to replace the class-conflict/forces-relations of production model with what he calls a "work-interaction" model.)

Finally, the basic charge of suboptimization can be met on its own terms in later capitalism. State capitalism can easily claim to have solved that problem and to have regulated the cycles of early capitalism far more efficiently than other available historical models (certainly better than command economies).

In a radical extension of the scope of ideology critique, Horkheimer and Adorno connect the problem of technology to the "dialectic of enlightenment" in modernity in general, and so connect the mastery of nature to the mastery of others and to an attempted legitimation of domination, control, and psychological repression which, they maintain, is ultimately self-undermining and de-legitimating.[10] The best known form for such a critique (despite many differences) is Marcuse's *One-Dimensional Man*.[11]

In the account given by Horkheimer and Adorno in *The Dialectic of Enlightenment*, the prevalence of technology in modernity should be understood as a central aspect of the "positivity" or "identity thinking" characteristic of the Enlightenment scientific revolution itself, characteristic, even, of the appeal to rationality throughout the Western tradition. Whereas Marx had understood science and technology as progressive forces, helping to create the material conditions for capitalism's self-overcoming, Lukacs was the first to charge that science and technology also assume ideological functions in capitalism, contributing to an ideological distortion he called "reification." Horkheimer and Adorno (to a large extent developing a Nietzschean theme) radically extend this sort of critique.

What poses in modernity as a rationally enlightened attack on superstition, mythic consciousness, religion, and feudal social practices is presented by them as not only narrowing the arena of rational discourse (with great psychic costs, as in their studies of fascism), but as a form of thought incapable of, and deeply resistant to, self-critique, and as a way of linking rationality in the natural sciences and the social sphere with total control and predictability, in a way that again cannot assess or reflect on the ends served by such control. Incapable of such deeper reflection, Enlightenment thus itself becomes a myth or ideology, a promotion of control or power for its own sake, to the point of pathology. Horkheimer and Adorno do not treat this connection between rationality and domination as a historical phenomenon peculiar to capitalism and the predominance of the commodity form of labor power, as Lukacs does. It seems to them characteristic of all attempts at integrative rationality, as visible in the Odysseus and Sirens story as in Faust.[12] This will mean that the critical contrast with such objectification and domination will have to be a rather romantic notion of "the natural" and a proposed reconciliation with nature that is, in Kantian terms, precritical.[13]

Moreover, especially in Marcuse's account, such an ideology is far more successfully *integrative* than any previous one. Thanks mainly to the culture industry, what previous critics identified as signs of strain and potential contradiction in such integrative programs – the subjective experience of alienation, lack of reconciliation with others or with the system – have been eliminated. Individuals are progressively more reconciled, at a deeper, more psychologically complex, and perhaps more permanent level, to social authority.

There is a notorious problem with such accounts. The criticisms vary, depending on one's interpretation of the position, especially the extent to which one takes its proponents to be offering an indictment of the structure of modern scientific method and technology as such. An attack on the modern relation to nature (and others) as essentially a relation of domination, presumes that there are alternative models *in* the natural science tradition that could preserve the canons of objectivity, repeatability of experiment, testability of hypotheses, relatively clear confirmation relations between observation and theory, etc., but that do not embody the relation of domination. It presumes as well a model of technology not wedded to the notion of mastery of nature. It is not at all clear what a New Science or New Technology would be like.[14] (It is also not clear to what extent Marcuse is committed to such a notion.[15]) In the case of Horkheimer and especially Adorno, something like this problem produced the wholly "negative" notion of resistance so associated with their program. Simply resisting the transformation of social relations into managed, technically modeled, or bargaining relations, and of natural-aesthetic relations into manipulative, means-ends relations, seems to be touted as an end in itself. For many, such a conclusion reduces resistance to little more than a symbolic gesture.

This is of course not the last word on the approach suggested by Horkheimer, Adorno, and Marcuse. Worthy of note is a recent book by Andrew Feenberg, *The Critical Theory of Technology*, which attempts to revive some of Marcuse's insights without the unacceptable utopianism and romanticism Marcuse appeared committed to.[16] This involves showing the ways in which the design, implementation and organization of technology are, in various ways, historically contingent; that they, at least partly, reflect the interests of "elites" who do the designing and implementing; and that some form of a democratization of the work force can make the best, most just, social use of the now critically revealed "contingencies" in design and implementation.

None of this, he argues, requires the familiar "trade-off" between efficiency and justice sometime said to be at stake in traditional debates. This approach so historicizes the question of technology itself that there is not such thing as, simply, "technology," or *the* technological en-framing; there is technology designed in a certain social period for various tasks, embodying various ends, organized under certain normative assumptions.

The "critical theory of technology" promoted by Feenberg has a number of virtues. It avoids the limitations of the instrumentalist account of technology (a tool is just a tool, a hammer whether used in carpentry or to bang a tree trunk in Samoa) and the excesses of the "substantialist" approach, wherein technology embodies a world-orientation. Against the former it points out the variety of contingent ways a particular technology for a particular purpose was designed, and how important political notions of administrative control and hierarchical principles of organization were inherent in such design. Against the latter it makes the same point about contingency: technologies represent complicated and often contradictory political decisions, even if it is still possible to maintain an essentially human interest in efficiency and productive power.

However, as Feenberg admits, there is no compelling reason to think that any sort of democratization of industrial organization and technology design will simply lead thereby to a substantially different form of production. It would certainly make such reform *possible*; there are good reasons to think that various reforms enhancing autonomy, the chances of being the "subject" rather than the "object" of workplace technology, reforms enhancing interesting, diversified work and in general self-respect, are made much more difficult by current imperatives of power and control and *not* by considerations of efficiency.

But the problem here is deeper and will help introduce the larger problems facing ideology critiques. *Whoever* is in charge of the design and implementation of technology will be an agent deeply socialized in a modern ethos. It is still not clear that such an ethos possesses the resources within it to sustain a political and ethical appeal to a reform that may result in a system just as efficient, but more humane and just. The pressure of the rewards now in place and the fragmented, often unclear basis of a call to reform and to the social solidarity and sacrifice needed to implement it may make it too risky an adventure for any modern agent. Democratizing may have a fair but dispiriting result: "relegitimation" rather than reform.

To some extent such a question is an empirical and/or a historical one, and the most a critical analysis can do is to point out misleading or ideological claims about the "necessary" constraints of efficiency, or the "requirements" of technological rationality. But the problem just suggested raises a much larger issue.

Aporiai in the "Technology as Ideology" Claim

While it is true that a massive social reliance on technology can blind one to various social, ethical, and even "ontological" implications of that reliance, none of the above accounts succeed in identifying what is fundamentally "unthought" in the ever increasing role of technology in modernization, nor what might be the implications of the changing social status of technology (our apparently declining confidence in the autonomy, or methodological purity, or even the very efficiency of the "purposive rationality" it embodies).

Rather, one needs to understand the original social appeal of potential technological mastery as a central aspect of the *ethos of the modern revolution itself*, a revolution that is not, I have argued elsewhere,[17] essentially a bourgeois, or capitalist, or scientific revolution. To make a very long story very short: the right sort of doubt to have about the nature of the social and cultural promise of technological mastery is not a doubt about a change in our fundamental ontological orientation, or about who is really and unfairly benefitting from the payoff of the promise, or who is benefitting from ossifying and reifying one stage in the historical development of technology, or whether we are becoming the objects of the forces we were the original subjects of, or whether a form of rationality has been thoughtlessly totalized, or whether technology might have been designed differently, in ways more responsive to the social needs of those who labor, etc. These are all important questions in their own right, but they are not, I think, the fundamental one.

If we make a few rather vague but relatively uncontroversial assumptions (at least for the sake of the present argument), the problem will have to be posed differently. Assume that there is a fundamental connection between the original justification of the modern revolution itself and the "mastery of nature" promise so essential to the contemporary influence of technology. To be sure, technological power can assume a different kind of importance in any number of different social situations: early industrialization, nineteenth-century American optimism, totalitarian regimes, etc. But any very general worry about the relation between technological power and such things as our understanding of nature, the nature of knowledge, or the possibility of democratic politics (the kind of things addressed by "ideology critiques") will require some attention to the uniquely modern understanding of the necessity for an ever expanding control over the forces of nature. There could have been such a technological orientation, or a supreme political and social significance to technological power, only when such mastery seemed both necessary and possible. Understanding the conditions under which that could occur requires, I shall try to suggest, a different sort of account than is presupposed in the standard versions of ideology critique.

So by contrast with the modern promotion of mastery,[18] the premodern emphasis on contemplation, the belief that the best regime was a matter of chance rather than human will, the insistence on an accommodation to natural *tele*, the traditional horror at the prospects of mass, collective action, are all assumptions that can be effectively countered only if the likes of Bacon and Descartes can successfully attack and undermine the bases of such claims, and then fulfill the promise of a comprehensive alternative vision, a secure, repeatable "method" capable (according to Descartes) even of challenging God's own words to Adam and of allowing us to "enjoy without any trouble the fruits of the earth and all good things which are to be found there."[19] Making this assumption means that if we want to understand the relative importance of growing technological power in modernity, its significance or meaning for modern societies, we need to understand the centrality of the technological promise to the possibility of a modern revolution, and so to the modern rejection of antiquity, and especially to the revolutionary notion that the future can be directed and controlled by human will. If this is so, challenging the role of technology in modernity, whether in terms of the straightforward political problems noted above, or in resistance to the orientation, conception of reason, alienating social relations, or false neutrality charged in ideology critiques, will require a reexamination of, assessment of, and alternative to *that* essentially modern imperative.

For example, in traditional accounts of the function of the "legislator" or statesman, a common assumption was that one function of those who held political power was to create a common ethical sensibility among citizens. The reproducibility of a society, its ability to rally support and fend off attack, to maintain its identity over time, required an extensive political project, judiciously and wisely administered by leaders and educators with unique talents. Individual *self-mastery* for the citizenry was assumed to be a primary goal of political (or politico-religious) life.

One way of asking about the emergence of the technological imperative is simply to ask about the fate of such a goal (a) when under the influence of Machiavelli, among others, such a policy comes to be seen as a wildly utopian, and political history reveals that (usually base) passions always guide human action, no matter which motives we wish would be determinative; or (b) under the influence of Hobbes, the authority of the legislator's claims to "know" what virtues, or manifestations of self-mastery, are most important to promote, is challenged when a devastating epistemological attack on the foundations of traditional politico-religious authority is mounted.

Under such conditions political life or collective action in general might come either to look impossible or possible only under radically altered expectations. The successful mastery of nature might finally make possible our being able to face the fact that a trustworthy self-mastery is simply impossible. "Unredeemed," hedonistic agents might still be able to secure common goods, however, as long as we (i) *change* our expectations about those securable goods – health, security, freedom from want, the chance for a commodious life, (ii) – and this is crucial – *are able to produce enough surplus to be able to appeal to such interests* – the *only* reliable social "glue" cementing us together – and be able to "pay off," and (iii) if we assume that our legislator need not transform or ethically educate the souls of the citizens, but can just efficiently *calculate what they will do*, and so "*manage*" well rather than "rule." The possibility of modern democracy, under the assumption that human beings are egoistic, passion-satisfying engines, would thus depend essentially on a qualitatively improved, "world-historical" leap in technological power. As often noted, if the technology of management is sophisticated enough, we might eventually not only come to expect and rely on such an egoistic, hedonistic conception of agents, but might promote and energetically encourage such activity, under the now familiar "private vices/public benefits" formula. (Or, ironically, an energetic technological optimism is required precisely because of a kind of philosophical pessimism, a great reduction in expectation about what sort of "guidance" philosophy might provide.)

Put a different, broader way, we need to note the fact that the kind of technical power that could make possible such a new politics is itself dependent on the successful promotion of a distinct, new sort of social ethos. In the most famous and disturbing case, the productive power necessary to generate the surplus that would make modern politics possible requires a culture of consumption and acquisition, indeed a culture of ever expanding, ever more "stimulated" consumerism. In such a context, the question of who *controls* the productive capacity, how its surplus is *distributed*, who *designs* the technology, etc., through all important issues, do not touch the fundamental problem. The links in the modernization process have to be taken in all at once: (i) the collapse of the premodern understanding of the connection between individual virtue and public life itself leads directly and unavoidably to (ii) the emergence of altered modern expectations about the narrow possibility of peaceful, coordinated activity, which in turns requires (iii) new, greatly expanded technological power, itself dependent on (iv) a socialization process, the "production of demand," that will itself decisively influence and constrain all modern political life. Once we understand the way in which the modern rejection of premodern politics was itself provoked by intellectual and social crises, an absolutely fundamental connection between such politics and productive or technological power, together with the "virtues" necessary to sustain it, comes into view and appears permanently to "frame" any possible account of the significance of any new steering or distribution program for the productive forces.

In this sense, the proper question of technology would be the question of modernity itself: is a distinctly modern epoch, one characterized by a radical attempt at a break with these sorts of traditional views and by the attempt to achieve true collective self-determination, possible? Without such an enterprise, however diverse and hard to characterize, there could have been no *centrality* to technological power in modern life.

By contrast, ideology critique seems inevitably linked either to some controversial account of origins (agents' true interests, hidden until "reflection" exposes them), or, as we shall see shortly, to the notion of a some sort of structural encroachment by one domain or "world" over another. The historical-social formation of those interests, especially the interest in autonomy, the control of destiny and of one's own body (originally but not exclusively embodied by property owners), and the historical sources for the growth of such "domains" (especially the historical reasons for the collapse of traditional, teleological worldviews, the emergence of "instrumentalist" models as the only publicly defensible notion of rationality), are, I claim, different and more important issues.

The right metaphor for understanding the extraordinary and potentially distorting appeal of technological power in modernity is not a hunt for hidden origins or a delineation of geographical boundaries but attention to the context, the historical moment when mastery in general would have seemed, with some historical urgency, an unavoidable *desideratum*. If we can understand the rising importance of technological mastery in this broader context, we will be less inclined to see that rise as some sort of Faustian bargain, prompted by hubris, narrow class interest, confusion about different domains of rationality, or as a lust for power. Ideology critique tends towards such explanations of *why* what is now claimed to be hidden or unnoticed got to be hidden or unnoticed, and though I cannot demonstrate the claim here, I want at least to suggest that such interpretations are implausible.

Still more simply put, the modern claim that the highest, publicly defensible good is a technically efficient mastery of nature – with all its implications for social organization, ethical relations, and public life – may be, in both historical and general terms, *rational*. Modern agents might not be confused by the implications of the "philosophy of the subject," their preferences might not have been wholly formed in a situation distorted by the influence of money and power, they may not have confused the ends of work and those of interaction, they may not be falsely universalizing or naturalizing a particular historical epoch. At least we should not beg any questions in making such claims, and that will require the broader assessment I am suggesting, an assessment of the resources within modernity for understanding the possible narrowness of such a conception of a rational end and the exclusion of others.

Such a broader view will also permit a more adequate understanding of the implications of the shifting social status of science and technology away from a privileged center, under the weight of various historicist, sociological, naturalist and other critiques, something not well understood in the rather jejune contemporary fascination with a possible postmodernism.

All of which seems to raise the stakes for understanding the role of science and technology in modern life to an unsatisfiable level (the old Hegelian "you've got to understand everything to understand anything" problem). In order to motivate this way of looking at things, I want at least to defend my claim about the limitations of the ideological approach and hope the alternative I have in mind will begin to emerge. My example will be Habermas's well-known argument.

Habermas on Technology as Ideology

According to Habermas's famous account, the great problem with technological modernization has little to do with technology itself, but with the way in which the influence, scope, and success of technology in modern life has tended to authorize only "purposive" or instrumental and strategic notions of rationality and to de-legitimate (as unresolvable and subjective) genuinely practical or political questions.[20] The imperatives of "work" – rational if efficient, efficient if productive – and "interaction" – rational if the norms of successful, genuine communication are realized – have become confused, and the specific form of rationality inherent in communicative action has been overwhelmed by the demands of technical efficiency. The "life world" has been "colonized," "steered" or encroached on by the demands of a self-regulating system.

This critique, which asserts that it is the totalization of the instrumental notion of rationality, its absorbing the categorially distinct forms of communicative rationality, that renders science and technology ideological, appears a sensible response to the historical limitations of Marx's original account, and appropriately cognizant of the undeniable benefits of technological power over nature and our own fate.

The account is largely motivated by a critical appropriation of Max Weber's original account of modernization as the progressive rationalization and demystification of various spheres or subsystems of modern life. On this account, control "from above," legitimated by appeal to cultural, religious, and mythic world views, constrains the development of purposive rationality in various spheres of life. The efficient satisfaction of basic needs, the creation of surplus, and hence leisure and luxury, are all impeded by such cultural (and basically irrational) constraints. This begins to change with capitalism and the creation of a purposive-rational system that demands its own continual expansion. Under the growing pressure of such expansion, traditional worldviews are transformed into private beliefs, incapable of functioning as universal or culturally stable forms of social authority, and the legitimacy of the capitalist claim to productive efficiency and universal satisfaction of interests wins out in a kind of competition for social power. The West is modernized.

This success, although a real advance when measured against the "systematically distorted communication" and the "fateful causality of dissociated symbols and suppressed motives" of premodern forms of communicative interaction, becomes itself repressive and ideological when it prevents any reestablishment of a genuinely interactive life among modern subjects, which is what Habermas claims happens, and why he claims this role for science and technology (in its presumption to be definitive of rationality as such) is ideological.

Habermas's approach generates the following problems, all of which return us to the general issues about modernization raised above. First, Habermas does not treat the extraordinary acceleration of technological progress in modernity or the modern reliance on technology as unique historical phenomena. Following Gehlen, he interprets them as extensions of the basic structure of all human purposive-rational action. We are simply getting much better at "aiding" *homo faber* in what he has always been interested in doing: moving about better, seeing and hearing better, producing and regulating energy, governing our actions more efficiently, etc. Work is a permanent, constitutive "human interest."[21]

This claim leaves unanswered the question of *why* work, or (to use Habermas's language) the imperatives of purposive-rational activity, became so much more in the modern age; why the ideal of mastery began to occupy a qualitatively different position on the social agenda.[22] Put in terms of Descartes's rhetoric, we need to know why (rather suddenly in historical terms) we should have turned so much of our energy to the "*mastery* of nature." That image suggests not merely an *extension* of our human interest in successful purposive action. Even viewed within the domain of work, of being able to get done what we want to get done (ignoring for the moment whether this involves a wholesale new relation to nature, or world-orientation, or understanding of Being), this image suggests a kind of urgency, a situation of insecurity requiring a military assault against an *enemy*, all not captured in Habermas's account.[23]

Secondly, Habermas supplements Weber's account of rationalization with his own picture of the competing requirements of work and communicative or interactive activity. He suggests a theory of modernity in which the premodern standards of communicative interaction, since they were prejudiced, distorted, products of repression, etc., "gave way" under the pressure of the successful expansion of productive capacity and the purposive-rational standards of rationality that go with it. It is unclear whether this account is limited to certain social and economic aspects of modernization, or is meant to identify the basic origin of modernization. The latter is suggested by the frequent references to Weber and to the larger intellectual, philosophical and religious issues supposedly called into question by the expansion of purposive standards of rationality into numerous subsystems.

Such an account, in the first place, downplays the philosophical crisis brewing in the tradition since nominalism. It would be hard to understand why the *promotion* of utility, i.e., our interest in becoming "masters and possessors of nature," should have displaced the notion of knowledge as contemplation or produced an account of knowledge (the "new way of ideas") subject to methodological rigor and control, unless such a philosophical tradition is taken into account.[24] For example, the importance of developments in technology could not possibly have risen to such a high spot on the social agenda without the essentially modern view that the source of most if not all human misery was *scarcity* (a new and quite controversial claim) and that scarcity was a solvable technical problem, nor without a new notion of "knowing as making," inspired

by developments in mathematics.[25] The simple emergence of a new, more optimistic view of how much of the basic scarcity problem could be technically
resolved would not be a deeply significant discovery had the classical view still
held sway: that the central political problem is an unimprovable finitude, a basic, permanent distinction between the few and the many, and the unpredictable,
wholly contingent congruence of wisdom and political power.

The metaphor of one view pressing on or pushing aside another is not very
persuasive. We need an account and an assessment of such a new view of the
basic problem. Why did it arise when it did? Was it a rational thing then to
believe? For whom? Under what conditions? The general picture that Habermas
paints, with one version of rationality in a kind of boundary war with another,
does not adequately account for the historical context in which they would have
first been seen as competitors, and in which their competition would have been
seen as significant.[26]

The essential point is this: without a sufficient understanding of that larger context, we shall be unable to understand the *consequences* of any sort of
acknowledgement of the "limitations" of instrumental rationality. It would not
matter if such limitations are claimed because of a growing tendency to historicize scientific and technological procedure, or to question the neutral, universalist pretensions of such procedures (by attention to sociological, psychological
or pragmatic origins), or as a result of a critical attack on the ideological totalization of purposive rationality. We shall not be in a position to understand
what such limitations amount to without quite a broad view of the historical
landscape.

If, for example, what Habermas calls the displacement of practical by technical questions occurred because *any* version of practical politics must appeal
to essentially premodern and no longer defensible notions of ends, teleology,
nature, etc., or can be resolved only be a kind of strategic bargaining among
agents with incommensurable goals, then there are no significant practical consequences from such claims about limitations. There would be no possible
agenda for the practical realm, and a growing lack of confidence in the standard
or traditional claims of science and technology to provide unprejudiced or neutral means to satisfy any sort of ends would result only in a greater skepticism
and social fragmentation, not emancipation.

Now Habermas claims in his own voice that there is such an alternative
agenda, one based on freeing communication from arbitrarily imposed limits
and distortions caused by the interests of money and power, and so promoting an
"ideal speech situation." This claim has given rise to objections that Habermas's
distinction between purposive and communicative rationality is often arbitrarily

drawn. On such a view, Habermas has a curiously positivistic understanding of "science and technology," as if they really are, if restricted to their proper sphere, as squeaky clean methodologically as traditionally maintained.[27] Many recent discussions in the philosophy of science, from historical and sociological studies of scientific practice to issues raised by a new generation of scientific realists, have created a number of doubts about the possibility of such boundaries, and at the very least raise as a possibility that there is no methodological way to isolate the purposive-rational dimension of science and technology and preserve it safely in its own domain. Admitting that the institution of science, its organization, hierarchy, criteria of success or of "good" or "central" as opposed to "bad" or "marginal" science, etc., and the design and implementation of technology, are everywhere already "symbolically mediated," that social values interpenetrate at every level, helping to define purposive effectiveness, need not mean we are committed to some contrary or new science. We may just have successfully pointed out the ever "embodied" nature of thought and the severe problems Habermas faces trying to keep things in their proper boxes.

Habermas faces a similar problem if it can be shown that the significance of the modern emphasis on the form of rationality embodied in science and technology cannot be explained as a result of some sort of contingent (and reversible) displacement of interactive by purposive-rational norms. If there is some more comprehensive historical context within which both the abandonment of traditional value systems and the allegiance to an instrumental notion of rationality could be explained and motivated, then promoting open allegiance to rules of ideal communicative equality within such a context or whole might, and likely would, simply institute a formally fair way of *relegitimating* the substantial anomie, fragmentation, and dissatisfactions of modernity, forcing us, after our freely arrived at and communicatively fair interaction, back to the narrow confines of strategic and instrumental rationality as the best concrete, realizable hope we have got for coordination.

Habermas of course disagrees and thinks, first, that there would be some sort of pragmatic contradiction in a situation where agents, even while conforming to ideal speech conditions, sanctioned instrumental social relations and power relationships justified only by instrumental efficiency, and second, that a very great deal in modern social life would change if such an ideal speech situation were achieved. His account is more sensitive to issues and more nuanced than I have been able to present it in this summary. In many ways it remains a powerfully critical approach to many aspects of modernization. But I do not think the basic strategy – the "separate into relevant spheres" approach – goes deep enough into the modern origins of such original separations, and so leaves too unclear the implications of the Habermasian delimitations. The most general and now quite familiar way to state these issues would be to say that Habermas and the whole ideology critique tradition still remain bound by the Kantian assumptions that made it possible, still wedded to the hope that a formal account of communicative practice, or the conditions of interest formation, or of reflexivity in general, will provide us with the critical tools necessary to understand modernity. My claim has been not that such a direction is misguided,[28] but that these issues need to be raised within a broader philosophic framework, one more sensitive to the substantive, historical, and practical issues at stake in the modern revolution itself, and thus more responsive to the claim that modernity and its technological implications may be, finally and decisively, "legitimate"; at the very least, in historical terms, "sufficiently rational."[29]

Notes

1. See Hannah Arendt, *The Human Condition* (Chicago: University of Chicago Press, 1958); Jürgen Habermas, *Strukturwandel der Öffentlichkeit* (Berlin: Leuchterhand, 1962).
2. See *inter alia*, the essays in Michel Foucault, *Power/Knowledge: Selected Interviews and Other Writings*, ed. C. Gordon (New York: Pantheon, 1980).
3. See the historical account in Norman Stockman, *Antipositivist Theories of the Sciences* (Dordrecht: Reidel, 1983), Section 3.2, "Critical Theory's Critique of Positivism: The Kantian Background,", pp. 43–51. Stockman's account should be supplemented by additional attention to the role of Lukacs. See Andrew Feenberg, *Lukacs, Marx and the Sources of Critical Theory* (Oxford: Oxford University Press, 1981).
4. See the three types of "pejorative" ideology critiques identified by Raymond Geuss in *The Idea of a Critical Theory: Habermas and the Frankfurt School* (Cambridge: Cambridge University Press, 1981), p. 13.
5. This theme appears in many of Heidegger's later writings, but I shall treat as typical such essays as "The Age of the World Picture," "The Word of Nietzsche: God is Dead," and "The Turning," in *The Question Concerning Technology and Other Essays* (New York: Harper and Row, 1977), and the concluding lectures in the Nietzsche series, *Nietzsche, Vol. IV. Nihilism*, trans. D. Krell (San Francisco: Harper and Row, 1982).
6. Jacques Ellul, *The Technological Society*, trans. J. Neugroschel (New York: Continuum, 1980).
7. Robert P. Pippin, *Modernism as a Philosophical Problem: On the Dissatisfactions of European High Culture* (Oxford: Basil Blackwell, 1991).
8. See the references above in n. 2, and the useful discussion in David Hoy, "Power, Repression, Progress: Foucault, Lukes, and the Frankfurt School," in *Foucault: A Critical Reader* (Oxford: Basil Blackwell, 1986), pp. 123–148.
9. For representative passages, see Karl Marx, *Capital*, ed. F. Engels, trans. Samuel Moore and Edward Aveling (New York: International Publishers, 1977), pp. 312–507, Vol. I, Part IV, "Production of relative Surplus Value," especially Chapter XV, "Machinery and Modern Industry." For the contrast between production under capitalism and after, see *idem, Grundrisse*, trans. Martin Nicolaus (New York: Vintage, 1973), p. 488.

10. M. Horkheimer and T. Adorno, *The Dialectic of Enlightenment* (New York: Seabury, 1972).

11. Herbert Marcuse, *One-Dimensional Man* (Boston: Beacon Press, 1964).

12. See Max Horkheimer, *Eclipse of Reason* (New York: Seabury, 1974), p. 176.

13. This is of course not true of the position developed by Adorno in *Negative Dialectics*, but that is another story. See my discussion in *Modernism as a Philosophical Problem*, pp. 151–156, and Habermas's statement of his differences with the Horkheimer-Adorno approach in *The Theory of Communicative Action*, trans. T. McCarthy (Boston: Beacon Press, 1984), Vol. I, IV.2, "The Critique of Instrumental Reason," pp. 366–399; and in *The Philosophical Discourse of Modernity*, trans. F. Lawrence (Cambridge, Mass.: MIT Press, 1987), Chap. V, pp. 106–130.

14. See Stockman, *Antipositivist Theories*, pp. 57–64 and 240–246.

15. Marcuse, *One-Dimensional Man*, p. 154.

16. Andrew Feenberg, *The Critical Theory of Technology* (Oxford: Oxford University Press, 1991).

17. Pippin, *Modernism as a Philosophical Problem*.

18. I am assuming that the modern enterprise does not merely seize the opportunity to extend a "natural," species-characteristic interest in the control of nature through labor and tools, but that the early moderns began to reformulate the range of natural events that could be mastered, what could count as such mastery (given the new influence of mathematics and a new attention to the problem of certainty), what such mastery was for, and the relation between such a goal and other desirable social ends. Given such a claim, understanding why such reformulations occurred cannot be answered by appeal to a mere extension of such a species-characteristic interest. I am thus disagreeing with, e.g., Habermas's account. See the discussion below.

19. Rene Descartes, *Discourse on Method* in *The Philosophical Works of Descartes*, trans. Elizabeth Haldane and G. R. T. Ross (Cambridge: Cambridge University Press, 1969), Vol. I, pp. 119–120. See *Genesis* 3:17.

20. Especially in Jürgen Habermas, "Technology and Science as Ideology," in *Toward a Rational Society*, trans. J. Shapiro (Boston: Beacon Press, 1970) and in Vol. II of *The Theory of Communicative Action*, trans. T. McCarthy (Boston: Beacon Press, 1987).

21. Habermas, "Technology and Science as Ideology," p. 87

22. See, for example, Habermas's account of how "pressure" from the development of productive forces brings about, as if by hydraulic force, the end of traditional societies in "Technology and Science as Ideology," p. 96.

23. Contrast the different account of the significance and unique characteristics of labor in the modern world in Ahrendt, *The Human Condition*, Chap. 3.

24. Among the many studies of this complicated intellectual development, see Hans Blumenberg, *The Legitimacy of the Modern Age*, trans. Robert Wallace (Cambridge, Mass.: MIT Press, 1983), and Amos Funkenstein, *Theology and the Scientific Imagination from the Middle Ages to the Seventeenth Century* (Princeton: Princeton University Press, 1986). See also my "Blumenberg and the Modernity Problem," *Review of Metaphysics* 40 (1987), 535–557, and Chap. 2 of *Modernism as a Philosophical Problem*.

25. See the account in David Lachterman, *The Ethics of Geometry: A Genealogy of Modernity* (New York: Routledge, 1989).

26. See the criticism of Habermas by Axel Honneth in *Kritik der Macht: Reflexionstufen einer kritischen Gesellschaftstheorie* (Frankfurt a.M.: Suhrkamp, 1985), and my "Hegel, Modernity and Habermas," *Monist* 74 (July 1991), 329–357.

27. See Stockman's account in *Antipositivist Theories*, pp. 109–112.

28. See the development of this theme in my "Marcuse on Hegel and Historicity," in R. Pippin,

A. Feenberg, and C. Webel (eds.), *Marcuse: Critical Theory and the Promise of Utopia* (London: Macmillan, 1988).

29. I am thinking here of Hans Blumenberg's strategy and terms. See his *The Legitimacy of the Modern Age.*

BELATED PESSIMISM: TECHNOLOGY AND
TWENTIETH-CENTURY GERMAN CONSERVATIVE
INTELLECTUALS

JEFFREY HERF
German Historical Institute
Washington, D.C.

Technological pessimism in the tradition of radical, that is, illiberal and antidemocratic German conservatism is less a product of the famous revolt against
modernity of the period of 1870 to 1945 than of the unconditional, unambiguous defeat of Nazi Germany in 1945. Technological pessimism on the German
intellectual Right is thus a more recent phenomenon than we often assume to be
the case when observers subsume it under the broader rubric of cultural pessimism, which indeed has a long historical trajectory in German history. Enthusiasm
for technological advance in twentieth-century German history was compatible
at times with the counter-Enlightenment tradition. Among former radicalized
right-wing intellectuals after 1945, a subdued and much tamed cultural pessimism coexisted with a mild technological pessimism. Not until the 1960s,
when cultural as well as technological pessimism reemerged on the political
and cultural Left in West Germany, did technological pessimism play a significant role in German politics.

Technological pessimism, as the editors of this volume note in their introduction, has been and remains part of the Western counter-Enlightenment.
It has been and remains more all encompassing than a criticism of particular
technological devices. From the antiindustrial critics of the eighteenth and nineteenth century, to the nationalist antimodernists of the first half of this century,
and then to the postmodernists of recent years, technological pessimists have
argued that the entire project of the domination of nature and the vision of a
good society resting on the growth of scientific and technological theory and
practice are flawed at their core. They have also argued that technology has an
autonomous direction impervious and often antithetical to human intervention
and intention, and that this autonomous technology is often the decisive causal
factor in historical development.[1]

But the belief in the autonomy of technology, and in its overpowering political impact has not been restricted to the ranks of technological pessimists. As I

argued several years ago in a study of "reactionary modernism" in Weimar and Nazi Germany, this reification of technology has also found advocates among enthusiasts of technological advance. Technological pessimism and reification of technology are not peculiarities of German history. All of the nations of Europe, the United States, and Japan, to name only the most technological- ly advanced societies, have produced cultural and philosophical traditions of pessimism and reification of technology. However distinctive Nazism was, fas- cism, and its associated authoritarian cult of technology, was a European-wide phenomenon. In the past decade, historians, especially in Germany and Great Britain, have argued that the German romantic revolt against the Enlightenment and its anti-Western illiberal traditions were also accompanied by very West- ern, very modern, and even liberal accomplishments in parliamentary politics, welfare measures, advances for women, extension of citizenship rights, and other developments associated with "the West," that is, Britain, France, and the United States.[2]

Yet, despite some important qualifications introduced in recent historical debates, the term *deutschen Sonderweg*, or a "German special path," retains important conceptual validity. This is due to the temporal simultaneity of crises of modernization in the late century. That is, late nineteenth-century Germany became a unified nation-state, an industrial society with a mixture of author- itarian and parliamentary institutions, and a modern economy all at the same time. The political and social crises that had been spread over four centuries in Britain and France were telescoped into fifty years in what Helmut Plessner aptly called *"die verspätete Nation,"* the "belated nation."[3] The conservative nationalism that emerged when the German state unified in 1870, and its mix- ture of fascination and revulsion with technology, became a defining feature of German identity and certainly of an anti-Western German conservatism until it was defeated in 1945.

Thomas Mann spoke of Germany's "romantic counter-revolution against the Enlightenment," while Georg Lukacs called Wilhelminian Germany the "classic land of irrationalism."[4] For much of the postwar era, historians of the intellectual and cultural origins of Nazism, such as George Mosse and Fritz Stern, tended to equate the German conservative political rejection of British and French political liberalism and cultural modernity with hostility to technology. They saw technological pessimism as an important part of German nationalism of the Right from 1870 through 1945, and National Socialism as motivated in part by a full-scale rejection of modernity, including modern technology.[5]

In my own work on what I called "reactionary modernism," I argued that the German intellectual and cultural Right in Weimar and the Third Reich rejected much of political and cultural modernity, including Enlightenment rationality, while it embraced modern technology.[6] Pessimistic about and hostile to the values of the Enlightenment and liberal politics, the reactionary modernists were anything but pessimistic about modern technology. On the contrary, they welcomed technological advance both as a practical necessity and as cultural redemption. The cultural accomplishment of these nationalist intellectuals and publicists of the post-World War I era Right was to have identified modern technology as a distinctively German creation, and thus as part of rather than alien to a nationalist revolt against universalizing forces of modernity. Where the antidemocratic cultural Right had spoken of technology or culture, the re-actionary modernists offered metaphors, key words, images, and philosophical terms stating that the nation could be technologically advanced yet true to its pre- and antimodern soul. German conservative and Nazi rejection of the ends of Western political and moral traditions, both premodern and modern, coex-isted with the incorporation of a potentially foreign phenomenon, technology, into the language and metaphors of anti-Western German nationalism.

The high priest of cultural pessimism after World War I, Oswald Spengler, called for a priesthood of engineers to establish a technologically advanced au-thoritarian state. Ernst Juenger (Jünger), in his essays and books of the 1920s, saw in technology a welcome authoritarian and totalitarian alternative to the fragmentation of bourgeois society, as well as a source of hope for Germany's future international regeneration. Carl Schmitt and Hans Freyer saw in technol-ogy under state control a welcome alternative to the domination of the economy over society. Werner Combart juxtaposed German productive technology with Jewish parasitic capitalism. Cultural representatives of German engineers be-fore and during the Third Reich associated technology with language drawn from German vitalist philosophies. They spoke of a Nietzschean will to power, and Schopenhauer's will to form as imminent in technology. They associated technology with an intrinsic aesthetic creativity, a clarity of form, and a use-value that they contrasted to the misuse of technology in an economy driven by exchange value. The Nazis incorporated many of these themes of the anticapital-ist intellectual Right into their propaganda of a technologically modern German racial state that had burst the fetters on technological development – fetters, they argued, imposed by the "Jewish" Weimar Republic. As Joseph Goebbels put it, National Socialism advocated the "*stahlernde Romantik*" of the twentieth century in contrast to the "soulless," antitechnological mood of the "bourgeois reaction."[7] In this way, portions of the Nazi leadership embraced technology even while trying to destroy Western civilization.

Not all of the intellectual and political Right, nor the entire Nazi movement, party, and government could be characterized as reactionary modernist. Exponents of traditional anti-technological views did find a place in the Nazi hierarchy. Racism did draw on antiurban, agrarian, preindustrial utopias. But enough of the leading intellectual and political figures of the movement, party, and regime embraced ideas similar to reactionary modernism to justify a revision of our view of Nazism as a movement driven by ideological hostility to technology, and to challenge the view, articulated by Ralf Dahrendorf and David Schonbaum, that there was a glaring conflict between Nazi ideology and practice in regard to the question of modernity.[8] Ideological reconciliations between enthusiasm for and critique of technology before 1933 offered support for a unity of Nazi ideology and technological advance after 1933.

The "reactionary modernism" thesis demonstrated that attitudes toward technology in modern German intellectual and cultural history are related to but not synonymous with cultural pessimism, and that modernity can be accepted or rejected in bits and pieces rather than as a totality. What Fritz Stern called "the politics of cultural despair," a generalized lament about modernity, was compatible with great enthusiasm for technology, albeit an enthusiasm grounded in irrationalist and anti-Enlightenment traditions. There is a second, and heretofore insufficiently noted, implication of the existence of a reactionary modernist tradition in German history. The rejection of Western political and moral modernity is part of the history of modern Germany since the emergence of a national consciousness in the seventeenth century. In the nineteenth- and twentieth-century history of German nationalist cultural pessimism, technological pessimism is, to a much greater degree than we have previously appreciated, a very recent historical occurrence. On the political Right in this century, it first became important only after 1945. This right-wing post-1945 pessimism was a faint echo of the anguished pessimism of the post-World War I era. It gave way to a chastened *Bildungsbürgertum*'s mood of political withdrawal and pleas for individual autonomy in the Adenauer era.

Yet technology pessimism did not disappear from German politics and culture after 1945. The reactionary modernist tradition, along with the rest of Nazi and other anti-Western nationalist ideological traditions, collapsed and ceased to exert a central impact on West German political culture, although various compromised individuals continued to hold positions of responsibility in government, economy, the civil service, and the universities. As the late West German scholar Richard Lowenthal put it, in West Germany from 1945 to the mid-1960s the German traditions of anti-Western cultural pessimism, romantic anticapitalism, discontent with modernity, and yearning for an integrated *Gemeinschaft* went into latency. In the 1960s, Lowenthal continued, there occurred a "return of the repressed," a "romantic relapse" in which these older traditions of cultural discontent reemerged.[9] However, they reemerged on the Left, not on the Right, denoting a fundamental shift of cultural politics. In the 1920s, the romantic anticapitalism of Georg Lukacs, Ernst Bloch, the Expressionists, and the Heideggerian Marxism of the young Herbert Marcuse were manifestations only of a small number of leftist intellectuals. The Communist and Social Democratic Parties were resolute modernizers. In the 1960s, the largely expanded intelligentsia in the new Left found the skepticism about technology, or "instrumental rationality," of the Western Marxists of the 1920s far more congenial than the confident technological optimism of the Social Democratic and Communist traditions. Technological pessimism and the lament about modernity returned to German politics but it did so on the Left rather than the Right.

These fundamental historical shifts – the end or decisive weakening of reactionary modernism on the Right and its replacement with a mild pessimism about technology, and the subsequent reemergence of technological pessimism on the Left in the 1960s – underscore the importance of 1945: the complete defeat of Nazism as a caesura in German history.

After the ambiguous outcome of World War I, leading figures of the intellectual Right looked to technological advance to reverse the war's outcome. Following the unambiguous defeat of World War II, the blend of cultural despair and a view of technology as a deliverance from national catastrophe ceased to play a central role for the intellectual Right. As Jerry Z. Muller has shown in his intellectual biography of Hans Freyer, some of the leading intellectuals drawn to Nazism, when faced with the collapse of their political utopias, underwent a process of "deradicalization."[10] In place of appeals for the totalitarian, integrated national community of the Weimar and Nazi years, some of the disillusioned intellectuals of the post-1945 Right stressed the inescapable complexity and differentiation of modern society, as well as the importance of a private sphere safe from the claims of the former stifling public community. For these postwar conservatives, the reactionary modernist cult of technology was linked to their earlier, radical, antidemocratic views. The crushing defeats of the German army, especially against the supposedly racially inferior peoples to the east, fostered disillusionment among many German soldiers with Hitler, Nazi racist ideology, and with the cult of "German technology."[11]

Reactionary modernism, the blend of cultural despair and enthusiasm for technological advance that followed World War I, was closely linked to the *Dolchstoßlegende*, the idea that Germany could have won the war had it not been stabbed in the back by the revolution of 1918–1919. The antidemocratic Right in the Weimar years attacked the Versailles Treaty's restrictions on German rearmament. If through a technological quick fix, or, as Ernst Juenger argued, if "fire and movement" could be reintroduced into stalemated German military tactics, then perhaps the defeat of 1918–1919 could be reversed. The total defeat of 1945, by contrast, made it apparent Germany's advanced technology could not compensate for the political, strategic, and moral failures of its leaders.

Nazi ideology and the reactionary modernist embrace of technology came to an end in 1945 among those around Adenauer's Christian Democracy who assumed leadership of right-of-center democratic politics. Among postwar conservative intellectuals, the irrationalist embrace of technology gave way to a subdued technological pessimism, one that included the belief that technology had an autonomous and negative power, which threatened individuals, peace among nations, and democracy. Where the intellectual Right had once hoped to transform the world with technology, after 1945 its leading figures retreated to cultivation of a private sphere safe from the "threats" they saw in a technological, mass society. Though they gave up their aspirations to political leadership, their post-1945 technological pessimism, intentionally or not, could serve as a highly cultured and humanistically tinted apologetics. If technology was the great danger to humanity, the moral decisions of individuals became less important, as if technology had been a cause of the German catastrophe. Technological pessimism after 1945 had another apologetic consequence in West Germany. Like all complaints about modernity in general, it abstracted from particular, historical events and features of nations and societies, in this instance, German history. Rather than explore German history in greater depth, technological pessimism focused on a generalized modernity, one with no particular link to German peculiarities.

I will now examine the mixture of change and continuity in the traditions of the intellectual Right in Germany after 1945 by examining works by Ernst Juenger, his brother Friedrich Georg Juenger, Arnold Gehlen, Martin Heidegger, and the Nürnberg testimony of Albert Speer.

Ernst Juenger was the leading literary exponent of an irrationalist embrace of technology after World War I. An opponent of the Weimar Republic, he advocated a new authoritarian, and at times totalitarian, regime, In the 1920s, Juenger the reactionary modernist viewed technology as a thing of beauty, an expression of a powerful and primal will to power, a key aspect of the masculine community of the trenches in World War I, and a symbol of a new and welcome authoritarian and "planetary" politics. Modern technology, in Juenger's view, was inherently in tune with the demands of a reviving German nationalism and of a truly modern world of total mobilization. After 1933, Juenger found the

Nazi mass movement too plebeian for his taste, and in the course of the 1930s, he quietly revised his views. In 1941, mourning the death of his son, he wrote *Der Friede* (The Peace), a collection of essayistic fragments condemning war and yearning for peace. After being circulated among those linked to the plot to assassinate Hitler, it was first published in 1945. The *"Rausch"* or intoxication with technology so apparent in his works of the 1920s, such as *Feuer und Blut* and *Der Arbeiter*, was gone. The abstract, generalizing references to vague but powerful forces characteristic of his earlier works continued in *Der Friede*, but the mood was sober and mournful rather than heroic and cold-hearted.

In contrast to his nationalistic writings of the 1920s, Juenger had discovered a common humanity. World War II, he wrote, was "the first general work of humanity."[12] The seeds of a true peace could come only from "the common good in human beings," specifically from "his best core, his most noble, self-less stratum."[13] Unlike World War I, World War II was marked by "greater heartlessness" on the part of those who fight for "ideas and pure doctrines" rather than for nation or country. He referred vaguely to "murderous technics" in unspecified areas in which men were hunted as wolves. The "theories of the previous century," whether of equality or inequality, now bore fruit in the war in which "thirst for blood" grew to unimaginable proportions.

Especially terrifying was the cold mechanics of persecution, of superior technology of decimation, the investigation and surveillance of victims by lists and files of a merciless police, which had grown to the size of armies. It seemed that all order, all discovery of the human spirit had been transformed into a tool of oppression.[14]

Juenger referred to innocents whose only crime was "their being, the stigma of birth. They fell as sons of their people, their fathers, their race, as hostages, as believers of inherited faiths of bearers of their convictions, which laws passed overnight turned into a stain."[15]

From this landscape of death loomed the names of the great residences of death, in which ... whole groups of people, whole races, whole cities were exterminated, and where the leaden tyranny in alliance with technology celebrated endless blood marriages (*Bluthochzeiten*). These caverns of murder will, for years to come remain in the memory of human beings; they are the real monuments of this war, as Douamont and Langemarck were earlier. They [Douamont and Langemarck] could arouse pride as well as suffering; here there remains only sadness and depression because the damage was of a sort that touches the human species, from which no one can escape guilt. Earlier, progress ended with its thought and ideas; the all too clever spirit of invention lead to this swamp.[16]

He wrote of "hordes" who "were driven like cows to be slaughtered to graves and crematoria" where they undressed, were "assassinated like sheep" and forced to dig their own graves.[17] Labor and science were placed in the service of death;

doctors killed the weak and ill instead of healing them. Where the unknown soldier had been at the center of suffering in World War I, now suffering was generalized and darker and reached "deeper into the strata of mothers."[18]

These references to tyranny and murder in 1941 and 1944 and to the veiled specifics of the Holocaust made clear Juenger's disgust with the Nazi regime. Considering that this work was written under a dictatorship, its elusiveness, the absence of specific nouns – Germans, Jews, Hitler, Himmler, the SS, etc. – is understandable. As in Juenger's writings of the Weimar years, events appear without subjects who bring them about. They simply happen. Victims are murdered, but beyond reference to "murderous technics" there appear no perpetrators or actors. While his enthusiasm about technology had dimmed, the reification of technology in his prose, his depiction of it as a phenomenon beyond human control, endured.

Juenger's view of war and peace was elusive, vague, at times maudlin and sentimental, far from the heroic realism of the 1920s.[19]

That this war must be won by all means that it must not be lost by anyone. Today it can be predicted that if it is not won by all it will be lost by all. The fate of the peoples has become closely intertwined. It has become inseparable. The peace will either lead to a higher order or a growing annihilation.[20]

Nor did Juenger clarify things with assertions such as "the war is decided against one another. The peace will be won with one another."[21] He did not go so far as to call for the defeat of Nazism.

Juenger did see the outlines of a new world order emerging, and technology would play a key role in bringing it about. In the "planetary" dimensions of the war itself, Juenger saw "our most secret will toward unity." Technology extended beyond the confines of old nation-states. The earth, which could now be overflown "in hours" and linked "in seconds" with "images, signals and orders," lay "like an apple in our hands." Human beings were assuming a new form (*Gestalt*) and taking the shape of "citizens of new states in wonderous forms."[22] Technical advances fostered a broader thinking. Steam engines, coal, railroads, and telegraph accompanied the development and unification of national states. Now electrical technology, motors, airplanes, radio, and the power of the atom made this world too narrow. Nation-states were too limited for the new forms of interchange made possible by technology.

Juenger moved away from German nationalism to Europeanism. Europe, he wrote, ought to become a "partner of the great imperiums," and move beyond nationalism.[23] The new order that Juenger saw on the horizon remained what he called the realization of "the life-form of the worker" (*der Lebensform der Arbeiter*).

In this regard, the peoples have become more similar and daily come more to resemble one another in so far as the total mobilization, which they have entered into is subjected to the same rhythms. This concerns not only armaments questions but also a deep seated process of transformation. The unleashing of the work process on the fronts is only one of its sides; the other, not visible but also no less consequential, plays itself out inside the peoples themselves. As a result, none of the nations will emerge from war in the same form in which they entered it.[24]

In the 1920s the worker was Juenger's symbol of the new man who heralded the marriage of technology, war, and dictatorship. Now, however, the worker was the new man who had the "cleverness and overview for a worldwide peace planning. He is the only one who already thinks in terms of continents and for whom planetary concepts and symbols are understandable. Hence he will advance the ferment of unification.[25] "... the heroic era of the worker, which was also its revolutionary era, will complete itself. The wild stream has dug its own bed in which it will become peaceful."[26] Juenger saw the model for such states "in which the most varied peoples, races, and languages are united" in Switzerland, the United States, the Soviet Union, and the English empire. "From these forms has crystallized a sum of political experience. It should be consulted."[27] He yearned for a geopolitical unity for Europe, which would have to combine unity and multiplicity.

Freedom meant freedom for multiplicity, for the differences of peoples, their language, customs, laws, education, art and religion. "There cannot be too many colors on the palatte."[28] He hoped for a diminution of competition of national states – he had in mind first of all Germany and France – so that in the "new house" one could be freer than in the old to be identified as from Bretagne, Poland, the Basque, Sardinia or Sicily.[29] From the celebrant of the unifying technological world, Juenger had become a spokesman for diversity, pluralism, and of identities that had been incorporated into the nation-states of the nineteenth century. Where Juenger the reactionary modernist had earlier incorporated technology into German nationalism, he now, in 1941 and 1945, saw technological development as fostering a postnationalist world. A key element of Juenger's thinking, especially evident in the 1933 work *Der Arbeiter*, was a celebration of raw power and of dictatorship, as well as repeated assertions that they were both natural complements of modern technology. In 1941, Juenger wrote that issues of space and borders had to be settled by alliances and treaties. The former celebrant of "total mobilization" now argued that total mobilization had made peoples more alike and more willing to accept the importance of law. He criticized nihilism and even spoke of a return to religion, human dignity, and human rights and of peaceful relations between free peoples.[30]

When he did mention technology, it was more often in connection with negative phenomena: the organization of mass hatreds, the bombing of German cities, and the extermination (he used the term *Ausrottung*) of unnamed "whole countries, and whole peoples." For Juenger, the technology of total war in World War II brought forth none of the aristocratic, masculine pathos of the *Fronterlebnis*. In the struggle between "the powers of annihilation (*Vernichtung*) and the powers of life," the "just soldiers stand shoulder to shoulder like the old nobility."[31] Gone were raptures about the aesthetics of destruction and the will to power evident in technology.

But reification remained. As in his previous writings on technological themes, *Der Friede* tends to view technology as the driving force of political developments. In *Der Arbeiter* or the essays about the glories of the front experience of World War I, Juenger argued that technology required authoritarian or totalitarian politics focused on the mobilization of national energies. In *Der Friede*, he saw it as a force that could bring about international understanding and a cosmopolitan, peaceful Europe. In both instances, his reflections were apolitical, in that they lacked discussion of the political causes of political events. Whether optimistic or pessimistic about technology, he consistently endowed it with political powers it did not have, and he consistently reified it. Reification and a deficiency of knowledge of or interest in the political causes of political events were a continuity in Juenger's reflections on technology before and after 1945. The military outcome of World War I and the legend of betrayal led Juenger to believe that a technological mobilization in a second world war would lead to a German triumph. The reality of the Nazi dictatorship and the impending signs of national catastrophe and defeat appeared to convince him that a technological fix would not solve Germany's problems after 1945.

Ernst Juenger's younger brother, Friedrich Georg, was also a writer of literary essays and fiction, also contributed to the "conservative revolution" of the 1920s, and also wrote cultural philosophical essays about contemporary problems. Among these was *Die Perfektion der Technik*, perhaps the most unambiguous denunciation of modern technology to come from a German conservative before the post-World War II era. He completed the text in 1939. It was first published in 1946 by Klostermann Verlag in Frankfurt am Main. In 1949 an English edition appeared under the title *The Failure of Technology: Perfection without Purpose*.[32]

Some of the chapter titles convey his animus against *Technik*. "The Invasion of Life by the Automaton," "The Invasion of Life by Dead Time," "The Victory of Dead Time over Life Time," "The Myth of Exact Science," "Socialism as the Surrender to Technology," "Technology Serves Not Mankind but Itself," "The Ravages of Functionalism," "Technology's Attack upon Law and Property," "The Stampeding of the Masses," and "The Downfall of the Mechanized State."[33]

His depiction of the negative impact of technology is unrelenting: machines devour the landscape, cities are "grotesquely hideous," technology "darkens the air with smoke, poisons the water, destroys the plants and animals."[34] Exploitation of factory workers "(about which socialism is indignant only so long as it is in the opposition) is an inevitable symptom of the universal exploitation to which technology subjects the whole earth from end to end." Unions and political parties only tie workers "closer to the progress of technology, mechanical work, and technical organization."[35] Clock time and traffic subject human beings to automation. To be sure, this was far from the menacingly cheerful celebration of technology that emanated from the propaganda organs of Nazi Germany in the 1930s. By 1939 the Nazis were claiming that the terrible effects of technology had been corrected by the National Socialist revolution of 1933. The official view of technology was anything but pessimistic and Goebbels himself went to great lengths to denounce technological pessimism as a legacy of "bourgeois reaction" which could not grasp the rhythms and "hot impulses" of the *stahlernde Romantik* of the twentieth century.[36]

Friedrich Juenger depicted, in Langdon Winner's term, an "autonomous technology," one possessing its own subjectivity and political requirements.[37] For example, "technological thinking is obviously collectivistic" and "presupposes an individual, freed and cleansed from all conflicting considerations, an individual that will abandon himself unreservedly to the collective."[38] Or, "in the field of law, technology not only turns against such individual rights as are still independent of its organization. It also turns against the right of association, the right of organization wherever the groups thus formed are contrary to its interests."[39] In Hitler's Germany in 1939, this opaque technological gloom could be a kind of *innere Immigration*, the intellectual's retreat from the enforced displays of public enthusiasm for Hitler at the apex of his popularity. On the other hand, Friedrich Juenger placed the entire weight of his criticism on technology, as if the existence of the Nazi regime was incidental to the erosion of the rule of law, individual rights, and the erosion of individual autonomy.

Perfektion der Technik did not contain an analysis of National Socialism. It did bemoan the emergence of the masses who were, for Juenger, a product of technical progress. Malleable, located in large cities, habituated to the mechanization of traffic, the masses lost even freedom of motion. "If we watch the passers-by on a moderately crowded street, we recognize immediately the mechanical and compulsory manner of their movements and attitudes, illustrative of how mechanical their life has become."[40] The masses, caught in tasks of a specialized nature, are susceptible to ideologies that fill a gap.[41] The mechanization of life, rather than taming or eliminating human instinctual urges, "intensifies the dark side of human nature."[42] "All the rationality of the technician cannot prevent the growth of a blind elementarism."[43] Yet while this is taking place, mechanization also grounds down individuals into a mass.

The mass is the foremost material for mechanization by technology. But to the extent in which the masses become subjected to rational organization, they become supercharged with blind elemental powers and bereft of all spiritual powers to oppose them. The masses are running berserk, now in blind furious enthusiasms, then again in a stampede of terrified panic that drives them irresistibly to hurl themselves blindly and madly into the abyss, just like cattle or lemmings. Those torrential dynamic forces which technology unleashes also sweep along the man in the streets who fancies technical progress to be his own. Technology spells the mobilization of everything which was immobile heretofore. Man too has become mobilized. He not only follows automatic motion without resistance; he even wants to accelerate it still more.[44]

Unlike National Socialism's fascination with technology as one of the "hot impulses of our time," in Goebbels' words, Friedrich Juenger placed technology in a conservative analysis of fear of the masses.[45] Overwhelming in its power, technology both created the raw material for Nazism's mass base while releasing blind, elemental impulses.[46] On the one hand, technological pessimism for Ernst and Friedrich Juenger was key to disillusionment and distancing from the cheering crowds of 1939. Their analysis was out of step with Nazi totalitarian enthusiasm about a totally mobilized society. It was a moment in the history of the deradicalization of German conservatism, and hence a contributing factor to a sobered, more democratic postwar West German Right. They had lost their appetite for another war of the unified *Volksgemeinschaft* without yet acquiring the ability to analyse the role of individuals, institutions, and groups in bringing about the disaster that was the background for their opaque expressions of pessimism and despair. The pessimism about individual autonomy in a technological era evident in Friedrich Juenger's *Die Perfektion der Technik* made for skepticism about mass participation in politics.

I want to add just a few words about Heidegger. Victor Farias has documented his enduring commitment to Nazism up to 1945, and his cold heart and bad memory afterwards. Theodor Adorno, and later Winfried Franzen and Hugo Ott, demonstrated the intimiate link between his philosophical views and his political actions in the Third Reich.[47] I have argued that his attraction to National Socialism was linked to his hopes that it would rescue the Germans from the devastations wrought by two thousand years of soulless, technological progress set in motion by the Greek stance of dominating nature.[48] Though his hopes were disappointed in this regard as technical advance continued under the Nazis, Heidegger, as Farias demonstrates, remained loyal to the lost cause. Michael Zimmerman has made a compelling case for a link between his views on technology and his engagement with Nazism.[49]

Heidegger's contribution to the German literature of technological pessimism was to argue that the political difference between dictatorship and democracy were insignificant compared with their shared technological advances. From this standpoint, the Americans and Russians were identical because they both fostered a "wild and endless race of unleashed technology and rootless organization of average individuals."[50] Only Germany, the nation in the middle, stood a chance of developing a new historical and spiritual force and by so doing could prevent Europe from being destroyed. Heidegger's great equation, his refusal or inability to make distinctions between democracy and dictatorship, did not end in 1945. After the war he concluded that the war and its outcome had not decided anything.[51] The two technological giants, having vanquished the Germans, now fought for total control of the earth. As Zimmerman points out, Heidegger argued that postwar Europe was still in the grip of the technological will to power which, he argued, was responsible for starting both wars. Ironically, this conservative, apologetic technological pessimism of the early 1950s, one that refused to make distinctions between dictatorship and democracy, shifted political coordinates from right to left and became an important theme of the antinuclear and ecological currents on the post-1960s Left. Though many important differences emerged as technological pessimism in Germany changed its political coloration, the impulse to see technology, not political decision making, as the source of modern problems, remained a powerful current of German thinking and feeling.

Among those conservatives and Nazis who spoke about technology, none had a larger audience or could speak on the basis of more experience with the Nazi regime than Albert Speer, Hitler's architect and armaments minister. His final speech at the Nürnberg trial in 1946 and his influential autobiography, *Inside the Third Reich*, were important documents of a turn toward technological pessimism in West Germany after 1945. More so than Ernst and Friedrich Juenger, Martin Heidegger, and the philosopher Arnold Gehlen, Speer did have something to say about politics as well technology.

In his final speech at the Nürnberg trial, Speer described Hitler's dictatorship as the first dictatorship of an industrial state in the age of modern technology, a dictatorship that employed to perfection the instruments of technology to dominate its own people. The radio and public-address systems, he said, had helped to subject eighty million persons

to the will of one individual. Telephone, teletype, and radio made it possible to transmit the commands of the highest levels directly to the lowest organs where because of their high authority they were executed uncritically. Thus many offices and squads received their evil commands in this direct manner. The instruments of technology made it possible to maintain a close watch over all citizens and to keep criminal operations shrouded in a high degree of secrecy ... Dictatorships of the past needed assistants of high quality in the lower ranks of the leadership also – men who could think and act independently. The authoritarian system in the age of technology can do without such men. The means of communication alone enable it to mechanize the work of the lower leadership. Thus the new type of uncritical receiver of orders is created.

We stood at the beginning of this development. The nightmare of many people, that peoples could be dominated by technology, was almost accomplished in Hitler's authoritarian system. Every state in the world now faces the danger of being terrorized by technology. This seems to me unavoidable in a modern dictatorship.

Thus, the more technological the world becomes, the more necessary as a counterweight is the demand for individual freedom and self-consciousness of individuals.[52]

The "criminal events of those years," Speer wrote, were not only an outgrowth of Hitler's personality. Their extent "was also due to the fact that Hitler was the first to be able to employ the instruments of technology to multiply crime." Speer warned of the dangers of "unrestricted rule together with the power of technology." Hence, the danger he saw lay in the combination of technology with dictatorship and the possibility it created for exponential multiplication of the power of the dictator's will. In his memoirs, published in 1969, he wrote that

The nightmare shared by many people, that some day the nations of the world may be dominated by technology – that nightmare was very nearly made a reality under Hitler's authoritarian system. Every country in the world today faces the danger of being terrorized by technology; but in a modern dictatorship this seems to me to be unavoidable. Therefore, the more technological the world becomes, the more essential will be the demand for individual freedom and the self-awareness of the individual human being as a counterpoise to technology.[53]

Unlike the pessimism of the conservative intellectuals, Speer's statement in Nürnberg mentioned the fact of political dictatorship. He did not argue that evil was inherent in technology or was its inevitable outcome. Rather he focused on the dangers of technology placed in the wrong hands. But had technology made those who followed Hitler's orders less independent of mind than subordinates of earlier dictatorships? Was the world of Nazi Germany one "dominated by technology?" Had not the Nazis used technology to serve their political purposes? To suggest that a more technological world would likely be a more dictatorial one was understandable in light of his own experience, but was this an adequate explanation of the crimes in the Third Reich? Did not such an explanation obscure human agency? As Richard Breitman's recent and excellent study of Heinrich Himmler reminds us, crimes were caused by criminals, not cogs in machines, whose ingenuity in mass murder rested on Nazi political convictions, not on the inherent impulse of the machine.[54] Modern technology had placed enormous destructive powers in the hands of evil men, but it had not made those men evil or made evil deeds inevitable.

The philosopher and sociologist Arnold Gehlen was the author of the best known work of technological pessimism in West Germany during the 1950s and 1960s, *Die Seele im technischen Zeitalter* (The Soul in the Age of Technology).[55] It was first published by Rowohlt Verlag in paperback in 1957 and sold 30,000 copies. By 1969, 90,000 copies had been sold.[56] Gehlen was born into the *Bildungsbürgertum* of Leipzig in 1904, studied philosophy, and completed his habilitation in 1931.[57] He joined the Nazi Party in 1933 and was active as an official of the party's organization for university professors.[58] When the Nazis forced Paul Tillich to resign, Gehlen was named as his replacement at the University of Frankfurt in the summer of 1933. After a short period as Hans Freyer's assistant in Leipzig, Gehlen was named to the chair of philosophy at Leipzig. In 1938 he moved on to a chair in Königsberg and in 1940 to another in Vienna. During the war he maintained his academic chair while serving simultaneously in an army psychology unit.[59] From 1933 to 1935 he had high hopes that the regime might be able to overcome what he regarded as the excessive subjectivism and individualism of the modern age. He set out to develop a proper, rather than primitive, philosophy for National Socialism. After being criticized by Nazi ideologist Ernst Krieck in 1935, Gehlen backed off from immediate questions. In 1940, he published his most important work, *Der Mensch, seine Natur und seine Stellung in der Welt* (Man, His Nature, and His Place in the World).[60] After 1945, Gehlen's Nazi affiliations impeded his professional career.[61] It was not until 1961 that he found employment at a university. He wrote very little about his past, tended to skip over the National Socialist period in his writings, and altered the texts of the Nazi era to eliminate passages that betrayed his affinities to Nazism. Nevertheless, Gehlen's views on technology influenced both West German conservatives and such independent leftists as Jürgen Habermas and the reformist communist Wolfgang Harich.

After 1945, Gehlen's attitude towards science, technology, and the division of labor was one of ironic resignation. Science and technology, he argued, could not offer the sense of purpose, norms, and duty that characterized traditional and archaic societies. Thus they opened the door to "subjectivism" as the prevailing mood of modern culture. Gehlen's *Die Seele*, like Friedrich Georg Juenger's *Perfektion der Technik*, contains little discussion of specific technological developments and their impact on society, of political decisions in favor of or against one kind of technology, or economic analysis of the costs of various technologies. Instead he developed a theory of technological development based on invariant features of human beings, the "means whereby man puts nature to his own service, by identifying nature's properties and laws in order to exploit them and to control their interaction." Technology was thus "part and parcel of man's very essence,"[62] The "replacement of organic by the inorganic" evident in new materials and new forms of energy is one of the most significant elements in the development of culture. Technique operates from the beginning "from motives that possess the force of unconscious, vital drives . . . "[63] There was no inherent pessimism or optimism in this message.

Gehlen described the growth of habitual, reified actions and modes of thought in modern societies "which resist criticism and are immune to objections." He referred to the "automatisms of consciousness," to formulas that operate "consciously yet unthinkingly," to a realm of moral issues that "must be handled in an automatic or semiautomatic manner if they are to be confronted with perturbed reflection."[64] "Let us visualize," he wrote, a society of "thoroughly reified and depersonalized 'functioning' individual(s)." They seek to minimize social friction, adapt a conventional mode of conduct "that is neither novel nor personal" and will thus fit the "requirements" of "systems of reference" and the "division of labor."[65] Like David Riesman's *Lonely Crowd*, Gehlen's *Die Seele* criticized the erosion of individual distinctiveness.[66] He warned that there was "a danger that the pervasive bureaucratization and reification of society would have the same effects as the mechanization of manual labor."[67] Gehlen quoted Alfred Weber, Ortega y Gasset, and Marx on the stultifying effects of he division of labor and concluded his section on "automatism" by asserting that the "specialized functionary, endowed with the specialized higher training so quickly acquired today, provides no defense against the relapse into barbarism."[68]

Die Seele im technischen Zeitalter ends with a paean to the merits of "personality," which were to be found in energy and inventiveness in everyday life in the service of higher ideals.[69] Gehlen was neither a reactionary modernist nor a technological pessimist in the Heideggerian mold; his plea to cultivate individual distinctiveness in the face of specialization and conformism was not a plea for political engagement in the service of democracy. The message of "the soul in the age of technology" was that of the traditional *Bildungsbürgertum*, that is, to cultivate one's soul and personality while remaining distant from the political issues of the day. In his conceptual abstinence, in refusing to offer an explanation of anything – politics, history, ideology, war and peace – through the single prism of technology, Gehlen's *Die Seele im technischen Zeitalter* stressed the complexity of the world and abandoned the simple and sweeping enthusiasm of both reactionary modernism or technological pessimism. In place of the utopia of a unified *Volksgemeinschaft*, Gehlen and other West German conservatives stressed the complexity and differentiation of society[70]

Conclusion

The defeat of Nazism was central to the emergence of technological pessimism among German conservative intellectuals after 1945. Before 1945 the most prominent German conservative intellectuals addressing the question of technology were reactionary modernists and thus not technological pessimists. Only as Nazism in practice became a nightmare, and then was totally defeated in the war, did reactionary modernism give way to technological pessimism, and then to a return to a conservative *Kulturkritik*. This is a story of radicalization, disillusionment, deradicalization, and sobriety, a *Neue Sachlichkeit* after 1945. In place of the uninformed, dogmatic assertions of the reactionary modernists or technological pessimists, West German conservatives, with the exception of the forgetful and unrepentant Heidegger, turned to more modest, more modern, and more complex ideas. Gehlen's *Die Seele im technischen Zeitalter*, while bereft of political insight into the German past and present, was far removed from the ecstasy and despair about technology of Ernst and Friedrich Juenger in the 1920s and 1930s. The traditions of cultural despair that had brought Germany and the world so much suffering were discredited after 1945.

In twentieth-century German intellectual history, technological pessimism and cultural despair about modernity have gone hand in hand. Ernst Juenger, Friedrich Georg Juenger, and Martin Heidegger all fall within the apolitical tradition of the *Bildungsbürgertum*, a tradition of aristocratic disdain for the give and take of parliamentary politics. They believed that technology was the primary source of deliverance or disaster. Gradually, first in parts of Speer's Nürnberg testimony, and in Gehlen's pleas for individual autonomy in the 1950s, the postwar intellectual Right learned that technology could not explain politics and society. To be sure, remnants of the idea that technology was autonomous or inherently attuned to particular kinds of political views remained, as did suggestions that technology, not individual human decisions, was responsible for the disasters of Nazism. But the grand enthusiasm, the belief that technology would serve a desired national community, had collapsed along with the collapse of the German dictatorship.

Cultural pessimism and discontent with Western modernity was a central feature of German nationalism and its response to rapid modernization in the nineteenth century. Technological pessimism among Germany's intellectual Right, however, was a more recent historical phenomenon than Germany's discontent with modernity. It was a direct result of the reality, and then the defeat, of Nazism.

Notes

1. On this tradition see Langdon Winner, *Autonomous Technology: Technics-out-of-Control as a Theme of Political Thought* (Cambridge, Mass.: MIT Press, 1977).
2. See Thomas Nipperdey, "Wehler's Kaiserreich. Eine kritische Auseinandersetzung," in *Gesellschaft, Kultur, Theorie: Gesammelte Aufsätze zur neueren Geschichte* (Göttingen, 1976), pp. 360–389; and his *Nachdenken über deutsche Geschichte* (Munich: C. H. Beck Verlag, 1986), especially "Probleme der Modernisierung in Deutschland," and "War die Wilhelminiische Gesellschaft eine Untertan Gesellschaft," pp. 44–59 and 172–186; David Blacbourn and Geoff Eley, *Peculiarities of German History* (New York and London: Oxford University Press, 1984); and Jürgen Kocka, "German History Before Hitler: The Debate about the German *Sonderweg*," *Journal of Contemporary History* 23 (1988), 3–16.
3. Helmut Plessner, *Die verspätete Nation*, (Frankfurt/Main: Suhrkamp, 1974).
4. Thomas Mann, "Deutschland und die Deutschen," in *Thomas Mann: Essays*, Vol. 2, *Politik*, ed. Herman Kunke (Frankfurt/Main, 1977), pp. 297–298; Georg Lukacs, *Die Zerstörung der Vernunft* (Darmstadt, 1962).
5. For example, see George Mosse, *The Crisis of German Ideology: Intellectual Origins of the Third Reich* (New York: Grosset & Dunlap, 1964); and Fritz Stern, *The Politics of Cultural Despair* (Berkeley: University of California Press, New York, 1961).
6. Jeffrey Herf, *Reactionary Modernism: Technology, Culture and Politics in Weimar and the Third Reich* (New York and London: Cambridge University Press, 1984).
7. For Goebbels' statement on the meaning of *stahlernde Romantik*, see Herf, *Reactionary Modernism*, pp. 195–196.
8. Rahlf Dahrendorf, *Society and Democracy in Germany* (New York: Doubleday, 1967); David Schoenbaum, *Hitler's Social Revolution: Class and Status in Nazi Germany, 1933–1939* (New York: Doubleday, 1966).
9. Richard Lowenthal, *Gesellschaftswandel und Kulturkrise. Zukunftsprobleme der westlichen Demokratien* (Frankfurt/Main: Fischer, 1979); idem, *Social Change and Cultural Crisis* (New York: Columbia University Press, 1984).
10. Jerry Z. Muller, *The Other God that Failed: Hans Freyer and the Deradicalization of German Conservatism* (Princeton: Princeton University Press, 1988).
11. See Martin Broszat, Klaus-Dieter Henke and Hans Woller, *Von Stalingrad zur Währungsreform* (Munich: Oldenbourg, 1988).
12. Ernst Juenger, *Der Friede* in *Betrachtungen zur Zeit* (Stuttgart: Ernst Klett, 1980), p. 195.
13. *Ibid.*, p. 196.

14. *Ibid.*, p. 201.
15. *Ibid.*, p. 202.
16. *Ibid.*, p. 202.
17. *Ibid.*, p. 203.
18. *Ibid.*, p. 206.
19. *Ibid.*, p. 210.
20. *Ibid.*, p. 210.
21. *Ibid.*, p. 211. [Der Krieg wird gegeneinander entschieden, der Friede will miteinander gewonnen sein.]
22. *Ibid.*, p. 212.
23. *Ibid.*, p. 213.
24. *Ibid.*, p. 221. In dieser Hinsicht sind die Völker sich sehr ähnlich geworden und gleichen sich täglich mehr an, insofern die Totale Mobilmachung, in die sie eingetreten sind, demselben grossen Rhytmus unterliegt. Es handelt sich hier nicht allein um Rüstungsfragen, sondern um tiefgreifende Umformungen. So bildet die Entladung dieses Arbeitsprozesses an den Fronten auch nur die eine seiner Seiten; die andere, unsichtbare, aber nicht minder wirkungsvolle spielt im Innern der Völker selbst. Auf diese Weise wird keine der Nationen aus dem Kriege in den gleichen Formen entlassen werden, in denen sie in ihn eingetreten ist. Er ist die grosse Schmiede der Völker, wie der die der Herzen ist.
25. *Ibid.*, p. 222.
26. *Ibid.*, p. 222. Der Friede ist dann gelungen, wenn die Kräfte, die der Totalen Mobilmachung gewidmet waren, zur Schöpfung frei werden. Damit wird das heroische Zeitalter des Arbeiters sich vollenden, das auch das revolutionäre war. Der wilde Strom hat sich das Bett gegraben, in dem er friedlich wird. Zugleich wird die Gestalt der Arbeiters, aus dem Titanischen sich wendend, neue Aspekte offenbaren: es wird sich zeigen, welches Verhältnis sie zur Überlieferung, zur Schöpfung, zum Glück, zur Religion besitzt.
27. *Ibid.*, p. 223.
28. *Ibid.*, p. 224.
29. *Ibid.*, p. 225.
30. *Ibid.*, p. 221. "nicht ihre innere Angelegenheit allein. Vielmehr strahlt jede Freiheitsminderung nach aussen, wo sie als Drohung sichtbar wird. Ebenso wie der Anspruch, am Raum and an den Gutern der Erde in gerechter Weise teilzuhaben, begrundet ist, so auch der Anspruch, dass die Rechte, die Freiheit und die Würde des Menschen geachtet werden, in welchem Landes es immer sei. Es kann kein Friede dauern als der, der zwischen freien Völkern geschlossen ist."
31. *Ibid.*, p. 236.
32. Friedrich Georg Juenger, *Die Perfektion der Technik* (Frankfurt am Main: Vittorio Klosterman, 1946); Friedrich Georg Juenger, *The Failure of Technology: Perfection without Purpose* (Chicago, Henry Regnery Co., 1949).
33. *Ibid.*, "Contents."
34. *Ibid.*, p. 22.
35. *Ibid.*
36. For Goebbels' comments along these lines, see Herf, *Reactionary Modernism*, pp. 195–197.
37. Langdon Winner, *Autonomous Technology: Technics-out-of-Control as a Theme in Political Thought* (Cambridge, Mass.: MIT Press, 1977).
38. *Ibid.*, p. 97.
39. *Ibid.*, pp. 97–99.
40. *Ibid.*, p. 127.
41. *Ibid.*, p. 129.

42. *Ibid.*, p. 143.

43. *Ibid.*

44. *Ibid.*, pp.144–145.

45. This fear of technology and "the masses" became an important part of European conservatism's response to and explanation of National Socialism. This variant of technological pessimism was not only a part of the conservative response, it is an essential component of Hannah Arendt's *The Origins of Totalitarianism*, in which Arendt argued that Nazism came about in part as an alliance of the elite and the mob. Juenger's assertion of a link between the mechanization of life and the growth of "blind elementarism" in both the individual and the mass shows striking parallels to Horkheimer and Adorno's discussions of "the revolt of nature" in *Dialektik der Aufklärung*. His work also evokes the prose of the Frankfurt School, with its critique of instrumental reason and comments on the cynical use of ideology by the culture industry.

Despite vastly different starting points, the right wing and the left wing of the German *Bildungsbürgertum*, the Juenger brothers, and the Frankfurt theorists, did indeed share considerable common ground and some common inspiration in the history of technological pessimism in Germany. They also shared an allergy to discussion of German history and politics. The basic thesis of the *Dialektic der Aufklärung*, after all, was that a number of extremely general causal factors, such as Western rationalism, the enlightenment, instrumental reason, etc., not the peculiarities of German history, were the keys to understanding Auschwitz. Perhaps one of the grounds for the popularity of "critical theory" in postwar West Germany was precisely its argument that it was not primarily in German history that the causes of Nazism and its crimes were to be found. This generalization of the causes of catastrophe, an allergy to proper nouns, political history, and German referents was a common element in the civilizational critique shared by Left and Right.

46. *Ibid.*, pp. 179–180. Friedrich Juenger, no less than Ernst, associated technology with men. "The sight of women employed in technical activities always has something incongruous about it. Lawrence rightly says that one leaves woman behind when one goes to the machine. And indeed, why should women be tinkering with machines? Their forte lies in quite another direction. Women pre-eminently belong to the life-giving side of existence, whereas the machines confront us with a dead world of sterile, sexless automatons. The machine is not animated matter like the golem of the Jewish saga. It is not clay enlived by a learned rabbi's magic, nor is it a man-made spirit, a homunculus. It is a dead automaton, a robot, untiringly and uniformly repeating the selfsame operation" (p. 180).

47. On this issue see Victor Farias, *Heidegger und der Nationalsozialismus* (Frankfurt/Main: Fischer Verlag, 1989). On Heidegger's ontology and his politics see Winfried Franzen, *Von der Existenzialontologie zur Seinsgeschichte: Eine Untersuchung über die Entwicklung der Philosophie Martin Heidegger's* (Meisenheim am Glan, 1975).

48. See Herf, *Reactionary Modernism*, pp. 109–111.

49. See Michael Zimmerman, *Heidegger's Confrontation with Modernity: Technology, Politics, Art* (Bloomington: Indiana University Press, 1990), especially Chaps. 3–6 for a most insightful and revealing discussion of Heidegger's wartime and postwar writings on technology and politics.

50. *Ibid.*, p. 28.

51. Martin Heidegger, *Was Heisst Denken?* (Tübingen: Max Niemeyer, 1954).

52. Albert Speer, *Inside the Third Reich*, trans. Richard and Clara Winston (New York: Avon Books, 1970), pp. 653–654; Albert Speer, *Der Prozess gegen die Hauptkriegsverbrecher vor dem Internationalen Militärgerichtshof, Nürnberg 14. November 1945-1, October 1946, Band XXI: Amtlicher Text in Deutscher Sprache* (München and Zürich: Delphin Verlag, 1984),

pp. 460–462. The original German reads as follows:

"Die Diktatur Hitlers unterschied sich in einem grundsätzlichen Punkt von allen geschichtlichen Vorgängern. Es war die erste Diktatur in dieser Zeit moderner Technik, eine Diktatur, die sich zur Beherrschung des eigenen Volkes der technischen Mittle in vollkommener Weise bediente.

Durch die Mittel der TEchnik, wie Rundfunk und Lautsprecher, wurde 80 Millionen Menschen das selbständige Denken genommen; sie konnten dadurch dem Willen eines einzelnen hörig gemacht werden. Telephon, Fernschreiber und Funk ermöglichten es, daß zum Beispiel Befehle höchster Instanzen unmittelbar bis in die untersten Gliederungen gegeben werden konnten, wo sie wegen ihrer hohen Autorität kritiklos durchgeführt wurden. Oder sie führten dazu, daß zahlreiche Dienststellen und Kommandos unmittelbar an die oberste Führung angeschlossen wurden, von der die direkt ihre unheimlichen Befehle erhielten. Oder siw hatten zur Folge eine weitverzweigte Überwachung der Staatsbürger und den hogen Grad der Grad der Geheimhaltung verbrecherischer Vorgänge.

Fü den Außenstehenden mag dieser Staatsapparat wie das scheinbar systemlose Gewirr der Kabel eine Telephonzentrale erscheinen; aber wie diese konnte er von einem Willen bedient und beherrscht werden. Frühere Diktaturen benötigten auch in der unteren Führung Mitarbeiter mit hohen Qualitäten, Männer, die selbständig denken und handeln konnten. Das autoritäre System in der Zeit der Technik kann hierauf verzichten, schon allein die Nachrichtenmittle befähigen es, die Arbeit der unteren Führung zu mechanisieren. Als Folge davon entsteht der neue Typ des kritiklosen Befehlsempfänger.

Wir waren erst am Beginn dieser Entwicklung. Der Alptraum vieler Menschen, daß einmal die Völker durch die Technik beherrscht werden könnten, er war im autoritären System Hitlers nahezu verwirklicht. In der Gefahr, von der Technik terrorisiert zu werden, steht heute jeder Staat der Welt. In einer modernen Diktatur scheint mir dies aber unvermeidlich zu sein. Daher: Je technischer die Welt wird, um so notwendiger ist als Gegengewicht die Förderung der individuallen Freiheit und des Selbstbewußtseins des einzelnen Menschen.

53. Speer, *Inside the Third Reich*, p. 654.

54. Richard Breitman, *The Architect of Genocide: Himmler and the Final Solution* (New York: Knopf, 1991).

55. Arnold Gehlen, *Die Seele im technischen Zeitalter: Sozialpsychologische Probleme in der industriellen Gesellschaft* (Hamburg: Rowohlt Taschenbuch Verlag, 1957). An English translation appeared in 1980 under the title *Man in the Age of Technology*, trans. Patricia Lipscomb, with a foreword by Peter L. Berger (New York: Columbia University Press, 1980). The replacement of "Soul" with "Man" in the title was an unfortunate decision for it displaces the work from a longer tradition of German reflection on the relationship between *Technik* and *Seele*, technics and the soul. Unless otherwise indicated, I will refer to the existing English edition. *Die Seele im technischen Zeitalter* was a revision of a shorter work, *Sozialpsychologische Probleme in der industriellen Gesellschaft* (Tübingen: J. C. B. Mohr (Paul Siebeck), 1949).

56. *Ibid.*

57. I have drawn the material in this section from Muller's *The Other God that Failed*, pp. 395–399. Muller's work must now be considered one of the most important for understanding the intellectual life of West Germany after 1945.

58. *Ibid.*, p. 396. See also Werner Rügemer, *Philosophsiche Anthropologie und Epochenkrise* (Cologne, 1979).

59. Muller, *The Other God that Failed*, p. 396.

60. *Ibid.*, p. 397. Muller notes that "in Gehlen's book it was the ability of institutions to provide their members with a firm and all-encompassing set of orientations that counted. It was the

stabilizing *function*, not the truth of falsity of the beliefs that legitimated institutions, that concerned him."

61. The American occupation authorities removed him from his professorship in Vienna, and his easily documented links to the Nazi regime made it impossible for him to get a job in the American zone of occupation.

62. *Ibid.*, p. 4.

63. *Ibid.*, p. 19.

64. *Ibid.*, pp. 145–146.

65. *Ibid.*, p. 147.

66. *Ibid.*, p. 147.

67. *Ibid.*, p. 152. Jerry Muller mentions that David Riesman's *Lonely Crowd* was "warmly received" by Arnold Gehlen, Hans Freyer, and Helmut Schelsky (p. 353).

68. *Ibid.*, p. 158.

69. *Ibid.*, p. 166.

70. For one of the most prominent debates in West Germany in the 1960s and 1970s over the issues of complexity, democracy, and technocratic thinking, see Jürgen Habermas and Niklas Luhmann, *Theorie der Gesellschaft oder Sozialtechnologie* (Frankfurt/Main: Suhrkamp, 1971).

TIME AND TECHNOLOGY IN HEIDEGGER'S THOUGHT

GABRIEL MOTZKIN

Hebrew University of Jerusalem

In recent years a spate of books has appeared on clocks, the measurement of time, chronology, and how the history of time measurement has affected our conception of history.[1] Quite another line of research into the past has investigated how the application of different time-frameworks discloses different durations and different intensities of change.[2]

This kind of historical investigation has usually been carried out without reference to the different philosophies of time current in the different philosophies of different periods. The merit of Stephen Kern's *Culture of Time and Space* lies in its attempt to combine time, technology, and cosmology in one account.[3] However, his analysis does not consider the relation between time and technology as seen from a point of view within a specific philosophy. In comparing philosophy and technology, he compares his summation of a philosophy to his depiction of the technology of the time. Analogously, analyses of a theory of time within a given philosophy rarely seek to link the consideration of technology to that of time,[4] or to the nonphilosophical discussions of time that were contemporary with the philosophy in question.

Martin Heidegger was a philosopher with an original theory of time, one that integrates subjective human time with the idea that all time is future time – an attitude based on Hermann Cohen's idea that future time is the primary mode of time.[5] However, Cohen set future time as the primary mode of time because he viewed the future as the mode of time that is basic to the process of scientific discovery, and hence to any theory that seeks to derive a theory of knowledge from the (imputed) procedures of science. Cohen's notion of the future in turn descended from Kant's notion of anticipations of experience.[6] Kant's concept of anticipations, however, despite its formulation in terms of experience, was not interpreted as a time that is itself experiential. Thus Kant's conception of anticipation was the source for most ideas about future time in subsequent philosophy, but it was detached from experience. Nineteenth-century psychologistic theories of consciousness instead sought to base the theory of consciousness' association of experiences on Kant's concept of intuition, thus implicitly accepting the idea that the experience of time is retrospective. Seen this way, Heidegger returned to a different Kant, one for which the a priori is itself an experience.

The originality of Heidegger's interpretation of technology is that it provides an original theory of the link between technology and science. In keeping with the notion that experience is itself an a priori structure and not an a posteriori one, Heidegger also believed that the experience of technology is ontologically prior to the elaboration of science. This understanding of technology was based on the idea that technology is a specific and given kind of human activity, one that is at our disposal in any culture or community. Such a characterization does not seem to be that different from empiricist histories of science, in which technology is also prior to science. However, not only the sense of priority but also the context of this priority is different, because experience is an a posteriori structure for empiricism, whereas the affinity between Heidegger and Kant can be found in their common view that experience is a priori. The question to be determined is what kind of experience: for Heidegger the aprioricity of experience does not necessarily mean that cognitive structures are a priori in the same way as experience. Hence the aprioricity of technology in experience must also be of a different kind than the aprioricity that would derive from empiricism.

If technology is a given human activity, then in the terms of traditional philosophy it would be an innate rather than an acquired propensity. Heidegger never used this kind of language because he would have rejected the dichotomy between innate and acquired as itself reflecting an artificial dichotomy between existence and experience. However, the question his own thought constantly confronted was whether this other propensity to create artificial dichotomies is not itself a given. In terms of the history of philosophy this same problem is whether the history of philosophy as it has occurred is a necessary or a contingent history. Despite his radical contingentism, Heidegger waffled when it came to the possible contingentism of the history of philosophy. He never adopted the Hegelian position that all the specific phases in the history of philosophy have been scenes in a necessary drama. Yet he never suggested that the philosophical quest could have been avoided, or indeed that it would have had a salutary effect on our experience had there never been a history of philosophy. On the contrary, he himself continued to be preoccupied with delineating the phases of this history. The history of philosophy could only have been avoided to the same degree that inauthenticity can be avoided: it can be avoided in specific situations but never as a general principle; it is ineluctable that most of human experience will be inauthentic experience. In our terms, then, the history of philosophy is contingent for specific human individuals, but it is not contingent for the history of being, i.e., for the history of the search for meaning. In that case, whatever value we attach to the productive or poetic ability and to that subset of it that we can call the technical ability, we must conclude that what is ultimately manipulable is our attitude to technology but not our use of it.

The question that underlies Heidegger's attitude towards technology is whether the adoption of a technological stance vis-a-vis the world must necessarily lead to an undesirable result, in other words whether the adoption of an inauthentic stance vis-a-vis the world must necessarily have negative consequences not only for the quest for meaning but also for physical existence. In our terms, however, the continuity of physical existence was not the principal question for Heidegger; he was much more preoccupied with what we would call the quality of experience. Heidegger attached no value to a purely physical existence: the cessation of the existence of the world or a nuclear holocaust was not a prospect that horrified him. In conformity with his position that all particulars can be determined in their existence only according to the contexts in which they can be understood, and that the central preoccupation of human being is the questioning investigation of these contexts, the loss of the possibility for questioning meaning worried him much more than the loss of existence, the loss of experience rather than existence. In that sense, being the victim of a genocide or a nuclear holocaust was not the ultimate evil in his view. Moreover, without knowing it we are already living in a condition of extremity, since we are in the process of losing the capacity of questioning the meaning of the context in which we are living. Many of these views affected both Hannah Arendt's vision of totalitarianism (the idea that a political regime could brainwash people) and Herbert Marcuse's idea of repressive tolerance (the idea that we can be brainwashed in such a way that we no longer understand our own experience).[7]

The misuse of technology cannot be transcended because it is a derivative of the will to power, and the will to power is a necessary concomitant of all concerned dealings with the world. However, such dealings with the world reveal a duality that is fundamental for the possibility of any further quest for meaning. This duality, while at the base of all inauthenticity, also provides the possibility of recognizing inauthenticity. Since the technological drive, like the will to power, cannot be transcended, it must necessarily create a condition of contrariety between what is produced and what gives it meaning. This contrariety is expressed as a contrariety between the time-structure inherent to human existence and the time-structure that is first revealed through human activity.

My thesis in this paper is both simple and circular. I believe that Heidegger's approach to the question of technology is a logical consequence of his concept of time, although one could question whether his analysis of technology was based in his analysis of time. Once he confronted the problem of technology, however, he could only do so in terms of the theory of time that he had already elaborated. Insofar as Heidegger's attitude towards technology was one of technological pessimism, the underlying question is whether or not Heidegger's position on time can be characterized as being one of temporal pessimism. It is then our task to define what could possibly be the meaning of a temporal pessimism as opposed to a temporal optimism.

On the other hand, when we look to the historical sources for Heidegger's conception of time, we have to conclude that Heidegger's conception of time is rooted in the technological conceptions of time that were current in his historical context. Logically, technology may depend on time; historically, the situation may be the reverse. We are obligated to a position for which logical and historical priority must dovetail, but the difficulties of an epistemology for which the two are inverse are notorious. One of the aims of Heidegger's thinking can be seen as an attempt to explain this paradox.

Usually, Heidegger's conception of time has been characterized as deriving from the Christian time of salvation and from dilation of time implicit in a kairological conception of time, of Divine presence modifying the linear order of time.[8] One difficulty with the Christian explanation is that it has been applied to all forms of thought in the post-Christian age, and thus minimizes the possibility of a break. Heidegger himself, in *Die Zeit des Weltbildes*, did take the position that the separation between modern secular culture and traditional Christian culture is minimal in a certain way.[9] However, he did not extend this continuity to include himself. In that essay he hinted that secularization involved the Christianization of the sensible world, i.e., the application of categories that were first devised for the intelligible world to the sensible world. Thus what really has disappeared is the traditional world of the senses, the traditional aesthetic, and not the world of the Divine or of ideas. Even this characterization, however, implies that a break has occurred; it is only the location of the break that has been shifted. However, it is not clear that this break involves a break in the conception of time; for Heidegger, the entire tradition has unfolded under the aegis of the metaphysics of presence. Hence the Christianization of the sensible world does not mean the abandonment of the metaphysics of presence. Moreover, the Christian conception of time was not a conception divergent from the tradition. The argument can be made that Heidegger sought to restore a Pauline conception of time against the tradition, but his use of that conception must be viewed as antireligious, a secularization of primitive Christianity.

The Christian explanation may be useful for explaining the survival of non-scientific conceptions of time in the scientific age, but it does not explain how a nonlinear conception of time has been modified by the development of a linear concept of time. Heidegger provided an explanation for the orientation to the present as resulting from the absorption and fascination with the world; but his explanation of the relation between the ontology of futurity and the metaphysics of presence was simplistic, because the structures he adduced were not formal structures of the kind that his concept of the future obligated him to elaborate, but rather material modes of behavior. He sought to show how feelings modify the experience of time, resulting in an inauthentic metaphysics and thus claiming that all of traditional metaphysics has been based on an anthropology that covers up its roots in intuitions. It is the conceptual sciences that are based on intuitions, while Heidegger's explanation aims beyond intuitions at providing a formal structure. Klaus Hartmann, in his paper, "The Logic of Eminent and Deficient Modes in Heidegger," has used a formal analysis of Heidegger's modes to show that his use of modalities is inconsistent.[10] Heidegger, in his explanations of the roots of inauthenticity and of what he viewed as the scientific illusion, was aware of the epistemological necessity of spelling out the logic of illusion. However, he tended to characterize that logic as proceeding from false premises and failing to show how the illusion is being constantly modified in relation to its object or end.

Another possible source for Heidegger's technological pessimism has been discerned in his nostalgia for the primitive and the primeval, which parallels his preference for pre-Socratic philosophy over all the mis-traditions deriving from Plato. But unlike Rousseau, Heidegger never believed in a preexisting harmony between nature and world; the disharmony between nature and world is innate. In one sense, this statement is misleading, for Heidegger argued that no distinction between subject and world can be drawn, that both are aspects of the same unity of meaning. However, insofar as we attempt to spell out the particular temporality of an individual human existence in the next world, that temporality always assumes as a working hypothesis a disharmony that is inherent in its own world between its own time and the other times that it confronts, even if these are only understood as subtimes of the world in which one is. Nature, whether a false or true category, is ultimately confronted as other. Certainly, Heidegger believed that primitive man had a better intuitive grasp of his interaction with the environment than modern man. In none of his characterizations, however, does he suggest that this grasp of the world derives from a unity or an identity between man and the world that is graspable as such by man. In contrast to the traditional position that the world is an irreducible other with man only knowing the world on one ontological level – i.e., at an epistemological level conformable to his own powers of knowing – Heidegger believed that subject and world are fundamentally the same, but that this identity is a limit-condition, so that it can never be grasped as such by man in the world, who requires ontological difference for his very existence in the world. The identity between subject and world could have been interpreted as implying an ontological identity pervading all of reality; that had been the Idealist position. Heidegger was constrained to explain the paradox of ontological difference in terms of the paradox of this identity between man and world. His recourse to a new characterization of time was his solution to this difficulty. However, the objection could be raised that the ontological difference is itself an epistemo-logical illusion, i.e., a trick whereby man masters the world by emphasizing the difference between himself and the world instead of emphasizing the identity. In that case, Heidegger would be an exponent of the same technological will to power that he so condemns. Only if it is grasped that Heidegger's conception of ontology is not the traditional conception of ontology, that it does not purport to deliver the secret of the *ens realissimum*, that it rather fulfills many of the functions of the traditional *epistemology* does Heidegger's position make sense. In his habilitation, Heidegger had praised the traditional distinction between the *modi cognoscendi*, *modi essendi*, and *modi significandi*. He made it clear that his own preoccupation was with the *modi significandi*.[11] In other words, the ontological difference between Being and entities is a difference in the kind of meaning that we attribute to the world, and not either a conceptual difference or a real difference. That conclusion assumes that a difference exists between meaning and concept, precisely Heidegger's position. Thus man's inability to grasp the world as such is not a consequence of the distinction between appear-ance and reality, but rather a consequence of the way in which man projects himself into the world. This self-projection can only be understood in terms of the particularity of human temporality. It is important to understand that for Heidegger, inauthenticity or its parallel in the world of knowledge – the rise of science – is not an ethical question, but a condition deriving from the fundamental paradox that man can never know his world as it is because of his very perspectivism, and that indeed the individual has always already sensed obscurely what there is to know about this perspectivism.

The objection could be made that it is unclear why Heidegger would think that we are in a position of "greater" alienation at one point in our trajectory than at another. His response must be the problematic one that he can discern what the ideal condition would be, even though he does not think that such a condition is attainable. The usual interpretation of Heidegger's account of the process of degeneration from a better intuition of being, which man once had, has been that it was caused by the invention of philosophy, modern technology being one consequence of this invention. According to this interpretation, the disharmony between man and nature is greater today than in the past. But this is not Heidegger's view. On the contrary, the concept of nature reflects the previous existence of this gap – the concept of nature implying our mistaking the world as our object, as if it were our object. In other words, our fundamental ontological mistake is a temporal one, caused by our sense of the temporal disharmony between ourselves and our object. It is the ascertainment of the existence of such a disharmony that motivates us to situate the transcendental object in either no-time or super-time, with the ultimate consequence that we then transpose this non- or super-time into our consideration of the cognitive subject. The immediate question is whether our efficient reason for the introduction of non-times is the greater manipulability of our object, or ourselves as our objects, by taking either as if the category of time is accidental to the existence of the object rather than a transcendental determination of its being. The remaining question is whether the introduction of non-time into one region of being (or meaning) implies an introduction of non-time into all considerations of meaning or being.

Heidegger was aware of the link between time and technology quite early in his career. In *Being and Time*, he refers to the intellectual history of his preoccupation with the question of the link between the technology of time and time itself.[12] He refers us to his lecture of 1915 on the concept of time in historical science, where he had already sought to understand the reason why chronological time was viewed as the primary way of understanding time.[13] His implicit point is that it is not the intuition of time as succession that makes chronology possible, but rather measurement and chronology that so penetrate our being that we intuit time as succession. His whole philosophy of time rests on the ideas that tense-time is prior to successive time, and that tense-time is not successive. Hence succession, and any system of measurement of time as succession implies a leveling of tense-time. Moreover, the reduction of tense-time to succession is not a direct reduction, but rather passes through several phases, for there is no inherent reason why tense-time must necessarily be reduced to succession. If succession could be applied in an unmediated manner to tense-time, that would be what we are measuring when we measure elapsed time as tense-time, and Heidegger did not believe that this is the case. Therefore a structuring of time must have occurred before it could be made amenable to measurement as succession. Heidegger viewed this middle stratum of time as historical time. Historical time is that time which has been transformed by human activity: man's activity in relation to his own time is to transform it into historical time. Once time has already been transformed by praxis into historical time, it then can be measured in a secondary operation. Thus what is being measured is a human experience of time, a material that has already been transformed so that it can be measured. This view has some affinities with Kant's idea of time as the form of inner intuition, for there, too, a transformation must have already occurred so that the temporal quality of appearances can be apperceived.

Chronological time assumes that what it is measuring is a succession, that the events in question already exist in an ordered succession. Otherwise, we would be forced to the position that ordering events in a chronology first emplaces them in a successive order; Leonardo could not have preceded Galileo until both have been located on the date line. If, however, we conclude that Leonardo preceded Galileo irrespective of our chronological system, then we must conclude that the experience of before and after is prior to the experience of time as chronology. The nub of the question is whether before and after have the same ontological character, i.e., if they can be ordered in a succession of events, since it is clear that chronology presumes succession. If we conclude that they cannot have the same character, because tense-time is not successive or continuous, so that not all the different meanings of before and after can be subsumed *tout court* under these terms, then we must conclude that there first occurs a process by which events are placed in a succession. This process is a human praxis, for which placing events in a succession makes them amenable to certain kinds of imputations of meaning. That may be a legitimate activity, but it requires a transformation of the experience of these events in such a way that their experience and their comprehension cannot be said to coincide.

Chronological time is then not a meta-time, a framework in which we embed our experiences, but rather a subset of historical time, where historical time is itself a mode of realization of the basic categories of human temporality. Historical time preexists both chronology and succession.

However, the ascertainment that succession is a transformation of tense-time and that chronology is a measurement of succession does not yet tell us why we should be interested in transforming succession into chronology. We have already reached the conclusion that a distinction exists between tense-time and succession. When we measure, however, we transform the succession of moments into a succession of a more determinate character, such as seconds, minutes and hours, where we assume that all these categories are predicated of the same substrate of succession, but where we have also investigated that succession in a particular mode of our conferring meaning on it, i.e., through measurement. It is this activity of measuring that can be called the technological activity, and it assumes that what we have already transformed into succession by the imposition of historical time on tense-time is nonetheless a primary given so that it can be measured as succession. When we measure time, we do not think that there is anything behind time that we are also measuring, nor do we think for the purposes of our measurement that time is a construct; for if we did, we should be obliged to devise instruments to measure that which is behind time. Technological measurement must assume that everything that can be measured has the same character of being amenable to measurement.

Heidegger thought that this problem of the measurement of time was a problem more of technology than of the investigation of time. He discerned a sharp division between time and its measurement, and hence concluded that the nature of time can never be investigated through the measurement of time. His explanation for the motivation behind the measurement of time was largely behavioral. I think, however, that even in his own terms, a further explanation of time-measurement must be given, for it is also clear that not all phenomena can be measured. Moreover, certain phenomena appear to require measurement more than others. In human history, clearly the first such phenomena were time and space. Once techniques of measuring them were established, the apparent difference between them only grew larger. However, Heidegger did not think that what we are measuring is time. Rather we create an entity that is amenable for measurement, in this case nature, which is then subdivided through time-measurement. Hence he claimed that the measurement of time as treated in the theory of relativity was not relevant for his kind of investigation, since the theory of relativity already presumes that we understand our being in time and how we are to go about investigating time in nature. Thus he implied in footnote iv. at H. 417 of *Being and Time* that the question must be one that would explain the problem of measurement in these terms.[14] I believe that still another problem arises here, for he assumes, as he already did in his lecture of 1915, that all kinds of measurement of time have the same ontological character – i.e., that the different kinds of measurement do not relate to different questions about the significance of existing in the world. In this sense, for Heidegger, not much difference exists between Newton and Einstein, since both set out to answer the same set of questions and both thought that the answers could be found by largely similar methods. Einstein's innovation, as characterized by Heidegger, was that "every clock 'has a history'" (H. 417),[15] i.e., that any instrument of measurement derives its ability to tell time because it has been made by man in accordance with his propensity to historicize.

This interpretation is reductive because it assumes what must be shown, namely that different measurements of time eventuate from the same set of questions; moreover, it also assumes that no question in a set of questions can breach the paradigm of questions in which it first was formulated. Therefore the game of significance is a closed-end game,[16] since the determination of the significance that I can find was already predetermined by the set of axioms I have presumed.

If we now combine two Heideggerian theories, the theory of the closed-end game and the noncoincidence of the logical and the historical, we can understand more precisely the origins of his pessimism. We already saw that Heidegger did not believe in the coincidence of the logical and the historical. However, this noncoincidence of the logical and the historical can be understood in different ways. We can infer that no connection exists between logical truth and historical contingency. In that case, however, we would be hard put to understand the nature of technology, since the production of any technology assumes that some connection between logic and history does exist. That last point is important insofar as it implies that the creation of machines with which to tell time affects either truth or human life or both, but it is only possible on the assumption that some kind of connection exists between truth and human life. Unless however we then can provide a third theory, we will be forced to choose between a strict historicism for which logic and history coincide at some point in their respective trajectories, and a logicism for which history is the progressively exact approximation to our conception of nature, a position close to Duhem's.

This dichotomy has assumed that we understand what we mean when we refer to the logical and the historical: i.e., that we understand that what we mean by logic is something such as a set of deductive truths about the structure of correct thinking, and by history, some kind of ordered succession of significant events that have occurred and are remembered by human beings as having occurred. Heidegger's point was that a reconceptualization of the category of the historical will obviate the issue of the link between the logical and the historical. Since the historical is a transformation of human time, one that is prior to the elaboration of succession as a principle for ordering that history, therefore the laws of the interaction between the logical and the historical do not follow the rules of succession. Once we have elaborated the temporality of the historical, we will be in a position to inquire after the link between the logical and the historical, the link that is essential for the understanding of technology. It is obvious that we can perform the same operation on the logical. Heidegger, however, did not perform this operation since he had no alternative logic to propose, and thus could only question the connection between logic and experience. However, he did have a proposal for an alternative reading of history, and it is in the implications of this alternative reading of history for the problem of technology that his interest lay. In other words, he ultimately, despite his own pronouncements, did not approach the problem of technology from the point of view of logic, bur rather from the point of view of history, i.e., from its effect in human history. It should be evident that such a view of technology is one for which technology itself is a tool more for the manipulation of history than for the manipulation of nature, that the importance of technology lies ultimately in its effects on human beings and not on nature, which, while a conclusion reached by many others, is not undebatable.

In the essay *Die Frage nach der Technik*, Heidegger applied the same categories of temporality to history that he had already applied to untransformed human existence.[17] Namely he now suggested that the historical process is itself teleological, a presupposition he had already entertained in 1915 when he had recommended the turn from a causal approach to history to a teleological one. A teleological approach to history, however, is one for which the essence is revealed at the end. The difference between such an approach and Hegel's lies in the relation of future to past: for Hegel, the future cannot be said to constitute the past. While history tends in a certain direction, it can only be understood after it has been experienced, through a retrospective turn. For Heidegger, the idea that the future constitutes the past, which is his notion about the essence of the phenomenon of becoming, means that this constitution must also be evident in history. His example is the relation of science to technology. While the technology of the industrial revolution only developed in the late eighteenth century,[18] and modern science developed in the seventeenth century, it would be a mistake to see the essence of technology as being the diffusion of the scientific worldview. On the contrary, technology, the "historically subsequent" is the "historically earlier." In this phraseology, Heidegger was playing with the double meanings in German both of earlier and of history. "Historisch," science, is "earlier" (früher); "geschichtlich," technology is "prior" (früher). But Heidegger's use of "earlier" was deliberately misleading, for he wanted to claim what is prior in essence is also prior in time. Thus what happens later is prior to what happens earlier.

Hence one must conclude that the temporality that is inherent in what we call logic is inverse to historical temporality; and the cross between the two lies rather in their revealing obverse aspects of the same phenomena, so that the apparent identity between the two is purely a matter of the perspectival position that we adopt, where, however, we have no way of privileging our perspectival position, since it too is subject to the same rules of the logic of illusion and the temporality of the historical process. Hence the real danger of technology lies in that it permits us to take the temporality of logic as the temporality of history, to make the presumption that there is some kind of ontological affinity between logic and history, and thereby to obviate the necessity of our quest for meaning.

The notion of inverse temporality allows Heidegger to suggest that history is governed by a logic of anticipation without espousing the determinism that would seem to be implicit in such a strictly teleological approach. He never suggests that the negative consequences must necessarily win out over the positive ones. His pessimism is a retrospective pessimism, not a prospective sense of impending apocalypse. History has gone badly because of the misuse of technology, a misuse that is implicit in the essence of the technological enterprise. On the other hand, technology could have positive implications if its significance in our culture were to change. This position is less tepid and less pious than it sounds, once Heidegger's recommendation for significance has been understood: he wishes to preserve the archaic significance of technology together with its modern development.

At the end of the essay on the question of technology, Heidegger recommends that technology be understood as a poetic production but that poetry should not be seen in the way we view works of art in our culture. Technology should itself be viewed as a possibility for the production of poetry, as itself a medium of constituting a relation to our world. I will not comment on the rigor or originality of this vision, but it does contain several interesting points. His model is his standard polarization of origin and end, a model in which the best way to approach the end is through the origin. The consciousness of the end is one for which the end is finite, so that a logic of the origin is not a logic of an open game, as it was for Cohen, but is rather a closed-end game. In this game, however, the time-relations are not linear. The same inverse relations that permit Heidegger to interpret one historical phenomenon as deriving its meaning from a subsequent historical phenomenon (which in fact may make sense for certain historical events) also permits the reimportation of the remote past, since we are no longer separated from it by historical distance. If history can only be understood in terms of the future constituting the past, then the constitution of the future by the past can only be understood in the future. A certain historicism is then a more salutary attitude in relation to the future than to the past. We should remember the past when we approach the future, not when we study the past, for which the future is the determinative category of time. In these terms, the essence of the Greek ideal of *techné* has not yet been discovered; the meaning of Greek historical experience can only be grasped when it has been reimplemented as a human practice in a different context, since all historical practice can only be grasped through reactualization. Heidegger has thus rescued the old ideal of history as education and example by separating it from any notion of the constancy of human nature or its applicability in studying the past. However, for this point of view the nihilistic disjunction between future and past could not be greater, since the inverse temporality of the one mode of time is also the inverse of the inverse temporality of the other mode of time. The past is modern and the future is archaic. The actuality of technology has been destructive, because technology has been used for the measurement (and control) of nature and not for the production of meaning.

It then would not be completely accurate to maintain that Heidegger is a technological pessimist. Technology cannot be transcended because we would have some other technology, perhaps a less damaging technology, if we did not have the one we have. But we would never be in a position of having no technology, because technology is part of the way in which we temporalize ourselves, i.e., part of the way in which we produce ourselves and time and attribute meaning to that self-production. However, it would also be wrong to conclude that Heidegger is a technological optimist, since he believes that the options for misuse in any human activity are always much greater than the chances for salvation, the dangers are always more oppressive than the salutary in the use of technology.

I have suggested in this paper that the interesting point about such an anodyne conclusion is not whether a German philosopher called Martin Heidegger thought this or that about technology, but the way in which he elaborated a theory in which technology is the premier example of the inverse relations between history and truth. In that spirit, it behooves us to ask the question that Heidegger avoided, the question of the relation between technology and logic, in terms of the ways in which technology affects or distorts our sense of what truth is, and how we could correct for such a distortion, if we could begin to specify it. Heidegger's point is that we could not do so either by designing an improved technology or by abandoning technology. The underlying assumption of his theory is that the quest for meaning is historical and that it is this quest that defines the time-relations in terms from which we can approach these questions.

Notes

1. David S. Landes, *Revolution in Time: Clocks and the Making of the Modern World* (Cambridge, Mass.: Belknap Press of Harvard University Press, 1983); Donald J. Wilcox, *The*

Measure of Times Past: Pre-Newtonian Chronologies and the Rhetoric of Relative Time (Chicago: University of Chicago Press, 1987).

2. Fernand Braudel, *La Méditerranée et le monde méditerranéen à l'époque de Philippe II* (Paris: Armand Colin, 1949).

3. Stephen Kern, *The Culture of Time and Space 1880–1918* (Cambridge, Mass.: Harvard University Press, 1983).

4. Françoise Dastur, *Heidegger et la question du temps* (Paris: Presses universitaires de France, 1990).

5. "In der Kategorie der Zeit haben wir die Anticipation der Zukunft erkannt, und zwar als das Eigenthümliche der Zeit." (Hermann Cohen, *Logik der reinen Erkenntnis* (Berlin: Bruno Cassirer, 1902), p. 391.) Also, "Nur Anticipationen ereignen sich in der Zeit; und nur correlative zu einem Vorwärts taucht ein Rückwärts auf." (*Ibid.*, p. 166.)

6. Immanuel Kant, *Kritik der reinen Vernunft*, trans. as *Immanuel Kant's Critique of Pure Reason*, trans. Norman Kemp Smith (London, 1929), pp. 201–208 (B207–B218).

7. Hannah Arendt, *The Origin of Totalitarianism* (New York: Harcourt Brace, 1951); Herbert Marcuse, "Repressive Tolerance," in Robert Paul Wolff, Barrington Moore Jr., Herbert Marcuse, *A Critique of Pure Tolerance* (Boston: Beacon Press, 1965, 1969), pp. 81–123.

8. Thomas Sheehan, "Heidegger's Early Years: Fragments for a Philosophical Biography," in Thomas Sheehan (ed.), *Heidegger: The Man and the Thinker* (Chicago: Precedent Publishing, 1981), pp. 3–20, especially p. 9.

9. Martin Heidegger, "Die Zeit des Weltbildes," in Martin Heidegger, *Holzwege* (Frankfurt am Main: Vittorio Klostermann, 1970), pp. 69–104. Trans. as: "The Age of the World-Picture" in Martin Heidegger, *The Question Concerning Technology and Other Essays*, trans. W. Lovett (New York: Harper and Row, 1981), pp. 115–154.

10. Klaus Hartmann, "The Logic of Eminent and Deficient Modes in Heidegger," *Journal of the British Society for Phenomenology* 5 (May 1974), 118–134.

11. Martin Heidegger, "Die Kategorien- und Bedeutungslehre des Duns Scotus," in Martin Heidegger, *Frühe Schriften* (Frankfurt am Main: Vittorio Klostermann, 1972), pp. 133–154, especially pp. 251, 269, 278.

12. Martin Heidegger, *Sein und Zeit*, 7th ed. (Tübingen: Max Niemeyer Verlag, 1953), p. 418, trans. as Martin Heidegger, *Being and Time*, trans. John Macquarrie and Edward Robinson (New York: Harper and Row, 1962), p. 499.

13. Martin Heidegger, "Der Zeitbegriff in der Geschichtswissenschaft," in Martin Heidegger, *Frühe Schriften*, pp. 355–375.

14. Martin Heidegger, *Sein und Zeit*, p. 417, *Being and Time*, p. 499.

15. *Ibid.*, p. 470.

16. Martin Heidegger, *Einleitung in die Philosophie* (unpubl. mss., 1929).

17. Martin Heidegger, "Die Frage nach der Technik," in Martin Heidegger, *Vorträge und Aufsätze* (Pfullingen: Gunther Neske, 1954), pp. 13–44, trans. as "The Question Concerning Technology," in Heidegger, *The Question Concerning Technology and Other Essays*, pp. 3–35.

18. *Ibid.*, p. 22.

THE POLITICS OF PESSIMISM: SCIENCE AND TECHNOLOGY
CIRCA 1968

EVERETT MENDELSOHN
Harvard University

When volume two of Lewis Mumford's *The Myth of the Machine*, entitled *The Pentagon of Power* was published in 1970, the critics were waiting and a number of reviews came rapidly.[1] It is almost as though the reviewers were anticipating Mumford's narrative and his pronouncements. On the ninth of November, 1970, Christopher Lehmann-Haupt, the veteran *New York Times* book-reviewer, opened with this description: "Lewis Mumford is perhaps our most distinguished flagellator." The secret of Mumford's success? "He's a megacritic, to coin a Mumfordian word." Lehmann-Haupt, always the judicious voice, was both bemused and annoyed by Mumford's latest effort, which, he pointed out, was his twenty-fourth book! He complained that there were too many self-citations in the bibliography, that there were too many self-congratulatory quotations from earlier works. (Both complaints seem, on the surface, to be justified.) That was not where Lehmann-Haupt left his judgment, however.

None of which, mind you, is meant to denigrate his major points, even if they don't quite justify the length of his new book and even if the language ... does not exactly scintillate. Quite the contrary, he comes as close to a precise definition of our modern ills as anyone that comes to mind.

Yet Lehmann-Haupt was uncomfortable with the book, almost certainly as much for its tone as for its composition. The theme that pervaded the text was Mumford's deeply pessimistic view of technology and science. This view was at the core of Mumford's sense of the fate of modern society and modern human beings. What emerged was a sharp and direct attack on the very roots of modernity. The problem of the present was traced directly to the past – the scientific revolution of the sixteenth and seventeenth centuries. The successful mechanization of the worldview achieved by Copernicus, Kepler, Galileo, and Newton had led to the clockwork universe, the clockwork society, and the clockwork human beings. As Lehmann-Haupt summarized it,

In the name of clockwork truth, technology was apotheosized. Mass production and human automation were the consequences, the modern American 'pentagon of power' the ultimate legacy.

The "pentagon" or five "P's" refer to Power as energy, Productivity for Profit, Political control, Publicity. What does Mumford find missing from this triumphant achievement? Human subjectivity and human values. Since these were neglected, Mumford believed, "so-called scientific truth was no more than superstition." Lehmann-Haupt provided a trenchant recapitulation of the Mumfordian critique:

The pentagon of power, with its missiles, computers, and nuclear warheads, is the modern equivalent of that ancient megamachine [the one that built the pyramids]: the scientists who tend them a new priesthood; and rocket flight to a lifeless moon our re-enactment of the worship of the dead.

Mumford's own "Preface" to *The Pentagon of Power* bristled with the language of the "wholesale miscarriages of megatechnics" and "collective obsessions and compulsions" that have led to misdirected and unsatisfying lives. And although the problems of the present drove his tone (1970 was a period of sharp confrontation and intellectual discontent in the United States), Mumford rightly claimed that the basic ideas of this book could be traced back to his classic *Technics and Civilization* (1934), composed during an earlier period of social crisis and discomfort with what technology had wrought.[2] At this very opening of his argument, Mumford alerted his readers to his long-standing commitment to the belief that "life itself [is] the primary phenomenon, and creativity, rather than 'conquest of nature,' [is] the ultimate criterion of man's biological and cultural success."

Lehmann-Haupt was not alone in his response to *Pentagon*; in fact among the formal reviews his may have been one of the most sympathetic. Lewis Coser, the doyen of the sociology of modern intellectuals, fairly sputtered with distrust and caricaturing. He labeled Mumford a " prophet of doom" (a title Mumford would have proudly owned) and saw him as someone who "hates almost all modern ideas and modern accomplishments without discrimination."[3] In those days of tense intellectual combat and literal battling on university campuses, Coser clearly placed Mumford on the "other side" of the barricades.

One of the sharpest and most fully drawn critiques of the *Pentagon* was written by Gerald Holton, Harvard physicist and historian of science. The tone and severity of the review he published in the *New York Times* is all the more surprising since Holton was part of the circle of friends that Mumford had built in Cambridge during his annual stay at Harvard and MIT in the

1960s. Further, Holton was a longtime believer in the necessity of scientists taking responsibility for scientific and technological achievements and, equally important for the intellectual politics of 1970, he was a visible critic of the Vietnam war.

Holton's review was given the most prominent display available in the *New York Times*, the front page of the Sunday book review section for December 13, 1970 (it continued with additional inner pages). Holton was most troubled by Mumford's sustained attack on the very roots of modern science, its commitment to objective knowledge and rationality. While Holton could be sympathetic with Mumford's bitterness about scientists' involvement in nuclear weapons design and production, he was fearful that *Pentagon* would become a potential weapon for those who wished to challenge rationality and science in general; indeed, that it would become part of what he perceived as the strong currents of antiintellectualism in the movements of "1968".[4] Holton's criticisms reflect one of the divisions in the Left-liberal groupings that opposed the United States' pursuit of the war in Vietnam. For Holton and many others in the scientific and scholarly communities it was not inbuilt errors or flaws of science that deserved to be challenged but rather the uses to which science was put and the willingness of scientists to be recruited to carry out those uses. Indeed, the very objectivism and rationalism that they found in science was, they believed, essential for the liberal critique of the government war politics of the time. Holton linked Mumford's attitudes with those of the youth who were in revolt and felt that Mumford had been caught up in the currents of the time: "he will now find himself at the head of a popular parade that may adopt his new book as a battle flag."

Holton was correct at one level. Mumford's words clearly resonated with views widely enough held at the time to gain very favorable response. The *New Yorker* was impressed enough with their sometime columnist's book to serialize it in four installments. William Shawn, the *New Yorker*'s editor, not only offered Mumford the handsome sum of $40,000 but in his letter to Mumford confided that,

... nothing we have done at the *New Yorker* has given me more satisfaction than our publishing 'The Megamachine.' It is a majestic and awesome work . . . And it may mark *the* turning point.[5]

Further confidence in the book and its public appeal was demonstrated by the Book-of-the-Month Club's offering of it as their main selection. By all measures it sold well. Mumford himself felt it was the best book he had ever written.[6]

Many in the established community of natural and social sciences objected to the *Pentagon* and its critique, but large segments of the literate and liberal public accepted the volume and seemed to applaud its challenges. Mumford himself claimed that he had not attacked science and scientists but rather that he had attacked "the Power Complex's threat to undermine all human values and purposes, including those of science itself."[7] Yaron Ezrahi in his recent study *The Descent of Icarus* examines in detail what others had sensed earlier – that the 1960s witnessed, among other things, an important breakdown in the unwritten contract that had linked science and society; since the early years of modern Europe, Ezrahi observes, science had been the source of the legitimating language of the modern democratic state.[8] Science had seemed to provide a space that would be free from chance, prejudice and arbitrariness; restraints would be imposed not by force but rather through "rational acknowledgement of restraint imposed by facts." Yet the very successes of modern science and its technologies had become a form of "blind power running wild."[9] Ezrahi makes another valuable observation about the very special relationship in the United States between science/technology and politics. It resulted in a decentralized civic order in which the citizen, not the expert, became the arbiter of knowledge. If he is right, and I believe he is, then the loss of the public faith in science and ultimately in science's ability to serve as the legitimator of the democratic state will involve a more profound critique of science than might be the case if only science's excesses or abuses were being challenged.

There is obviously a prehistory to Mumford's challenge in which segments of the public had prepared themselves to read and understand what Mumford was saying. But it is by no means a foregone conclusion when one looks at the variety of pieces of the historical record that the full pessimistic vision of the sciences would be adopted by any but a small minority. Indeed, there were still some very loud praises being sung to science even as Mumford himself was composing his criticism. In 1967, as the voices of dissent could be heard developing on the college campuses, Emmanuel Mesthene, the director of the then quite large ($5 million) IBM-funded Harvard University Program on Technology and Society presented his annual report, "Technology and Wisdom."[10] Its tone was joyful and celebratory.

Technology, in short, has come of age, not merely as a technical capability, but as a social phenomenon. We have the power to create new possibilities, we give ourselves more choices. With more choices, we have more opportunities. With more opportunities, we can have more freedom, and with more freedom we can be more human. That, I think is what is new about our age. We are recognizing that our technical prowess literally bursts with the promises of new freedom, enhanced human dignity, and unfettered aspiration.

While surely not planned as such these words could easily have formed the text for an IBM corporate advertisement. This scenario stands in sharp contrast to the dire prophecies of Lewis Mumford.

As though to serve as a counterpoint, just two years later and one mile down Massachusetts Avenue, on the campus of the Massachusetts Institute of Technology on March 4, 1969, the Union of Concerned Scientists convened a meeting specifically to challenge the current position of science. The statement issued as a call to the gathering proclaimed: "Misuse of scientific and technical knowledge presents a major threat to the existence of mankind."[11] The link to the war in Vietnam and the production of increasingly destructive weapons was made explicit. A call for concerted political action by scientists and engineers was part of the text which also included a request to all MIT colleagues (faculty and students) to stop their research activities for the day, March 4, 1969. Science – its present and future – were under intense scrutiny in the very citadel of established science and technology.

But what exploded in full public view through the events on university campuses during the academic year 1968–1969 had been some time in the making and only in retrospect do the several sources and trends appear unified. Lewis Mumford had tirelessly identified what he saw as the central theme in the critique of science and technology – Power, the "power system," the "power complex."

Perhaps it is unfair to bring forward the words of President Dwight D. Eisenhower, but they became so identified with the critique of power that their use seems legitimate, even if the source, the former war hero turned president, seems unlikely. For Eisenhower the words were no accident. On the eve of his departure from office, in his "Farewell Radio and Television Address to the American People, January 17, 1961," Eisenhower inserted himself into the debate quite consciously and I believe fully aware of his targets of choice.[12] Almost everyone remembers his warning against "the acquisition of unwarranted influence ... by the military industrial complex." The MIC became an all too convenient political code word in the following decades of antimilitary agitation. He went further, however, talking about the "technological revolution" that he believed fundamentally altered the industrial-military enterprise. Research was directed by the government. He characterized this research in consciously unflattering terms: "Today, the solitary inventor, tinkering in his shop, has been overshadowed by task forces of scientists in laboratories and testing fields." The "free university," which he believed had been "historically the fountain head of free ideas and scientific discovery has," he complained, "experienced a revolution in the conduct of research." Because of the high costs of this new form of scientific activity, government contracts have become "virtually a substitute for intellectual curiosity"; blackboards have been replaced by banks of electronic computers. There was an element of nostalgia in his fear that the nation's scholars would fall under the domination of government money, project allocations, and employment. But beyond the nostalgia for an older, perhaps libertarian view of the independent scientist-scholar, Eisenhower went on to express a much more direct fear; because of the changes wrought by this new revolution in technology and science, there was the "danger that public policy could ... become the captive of a scientific technological elite." The implied threat to democratic policy-making by an elite of experts and their allies in the new military-industrial complex provided the subtext of his address. Technical, military rationality, Eisenhower seemed to claim, was becoming the order of the day.

One reading of this episode is that the old soldier, turned president, had hoped in the closing days of his administration to achieve a treaty banning nuclear weapons tests. He had worked hard toward this end and although the negotiations had come close, newly powerful scientists (Edward Teller is the obvious figure) with their arcane arguments, aided by their friends in the military, thwarted Eisenhower's intentions. The response was Eisenhower's extraordinary step of issuing a public warning against the power of scientists.

There is no evidence that I have been able to locate to indicate whether or not Eisenhower was aware of the presence at Columbia University, during the brief interlude in which he served as the university's president, of the critical, progressive sociologist, C. Wright Mills. During these very years, however, Mills was ruminating and agitating about the *Power Elite* (1956).[13] He had pointed to the extensive links between science and the military, including military funding of science and the extent to which the general direction of basic scientific research was set by military considerations. In turn, he noted that theoretical military studies, strategy and policy had been deeply influenced by the involvements of the military with science. The "technical lieutenants of power" was the ironic sobriquet that Mills gave the scientists.

Just a little more than a year after Eisenhower's warning, in the summer of 1962, a voice from an entirely separate domain was raised extending the web of caution and doubt. Rachel Carson in *Silent Spring* opened her sharp attack on pesticides and the threat they held for the environment on a very somber note by way of a citation from E. B. White.

I am pessimistic about the human race because it is too ingenious for its own good. Our approach to nature is to beat it into submission. We would stand a better chance of survival if we accommodated ourselves to this planet and viewed it appreciatively instead of skeptically and dictatorially.[14]

Rachel Carson, known in the postwar decade as a popular and evocative nature writer, used the pages of *Silent Spring* to present an indictment of an uncaring, blinded industrial society that was despoiling its surroundings. She knew she could draw upon a sentiment deeply held in America's cultural tradition that valued the natural beauties of the environment. Her images and metaphors recruited the deeply held feelings that stretched from the writings of the conservationists of the nineteenth century to the mid-twentieth-century urban advocates of modern environmentalism and wilderness preservation.[15]

Had Carson stopped with the defense of the environment against the intrusions of pesticides she would have been successful enough. But she went further, picking up another sentiment rooted in the past but also newly enunciated in the present. She identified an attitude and outlook as the culprit: "The 'control of nature' is a phrase conceived in arrogance, born of the Neanderthal age of biology and philosophy, when it was supposed that nature exists for the convenience of man."[16]

By bringing the focus to the concepts of the control and domination of nature, Rachel Carson went beyond the simple criticism of science misused or misapplied to what she took to be the core outlook of the sciences, control and domination.

As though to assure readers that her critique of science and scientists was not merely an aside, she had on other occasions challenged the attitude of the "separateness of science" and the sense that it occupied some privileged position. At the celebration organized on her winning the National Book Award for *The Sea Around Us*, she shared her view. "Many people have commented with surprise on the fact that a work of science should have a larger popular audience"; but the notion that "science is something that belongs in a separate compartment of its own, apart from everyday life, is one I should like to challenge." For Carson it was impermissible to assume in the scientific age in which we live "that knowledge of science is the prerogative of only a small number of human beings isolated and priest-like in their laboratories."[17] The long held suspicion of science and scientists that informed significant segments of the conservationist literature, was restated by modern environmentalism's most popular author.

Silent Spring was published first in the *New Yorker* in three installments in the summer of 1962. It was a Book-of-the-Month Club main selection and also the choice of the Consumers Union Book Club. Over one hundred thousand copies were sold in the first four months after publication; it moved to the best-seller list at once where it was joined by a second Carson book, her republished

The Sea Around Us. By January 1964, 600,000 paperback copies had been sold.

Rachel Carson did not create the modern environmental movement but the decade in which *Silent Spring* was published, the 1960s, was a period of enormous environmental agitation. Recall that the decade had opened with the fierce battle over the dangers from atmospheric testing of nuclear weapons and that the first polluting substance mentioned by Rachel Carson in *Silent Spring* was radioactive fallout; indeed the pastoral image with which Carson opened her book was disturbed by a light ash-like substance that had floated down from the skies. Carson's message, critical of domination and control and scornful of experts and elites, provided an important, early and very clear voice particularly as it visibly came from "the gentle lady" as Senator Abraham Ribicoff referred to her in a eulogy delivered on the floor of the U.S. Senate at the time of her death in 1964.

I am not sure whether Eric Sevareid, the doyen of television's anchormen, realized what he was doing when he produced the "CBS Special" in 1962 in which key scientists from the pesticide producing chemical manufacturers and government laboratories were juxtaposed against Rachel Carson; their tone was knowing and arrogant against the backdrop of their laboratories and the lab coats, hers was reasonable and critical, she in a traditional, "lady-like" dress complete with its crocheted white collar.[18] Love of nature was pitted against science of nature.

Other voices of environmental concern, which became increasingly louder during the decade, were much more explicitly political. Theirs were the pronouncements that became the catechism of organized environmentalism; they explicitly crossed the boundaries between social, political and moral values on the one side, and scientific-technological thought and practice on the other. Barry Commoner, Paul Ehrlich et al. authored the texts that helped identify the ethos of the new movement.[19] Environmentalism, and those who responded to its message in the decade of the 1960s, represented a broadly popular, largely middle-class movement and audience who initially at least were politically moderate and socially middle-of-the-road. The texts they read and responded to related the story of nature – the environment – under siege, in part by the very elements that made theirs a successful industrial society. They heard as well a message critical of science as established and practiced.

There were other issues "out there" as well that reinforced the critique and also added to it explicitly political elements. The atomic bomb when used over Hiroshima and Nagasaki in 1945 was hailed by some as the "winning weapon" that proved decisive in ending the Second World War.[20] Yet even in the heady days of victory, others saw in the bomb and the very success of its explosive power a new threat which was laid directly at the feet of science. Lewis Mumford was among those who despaired at the inhumanity it demonstrated and the deep problem it foretold.[21] The lapse into the Cold War probably muffled the critical voices but they reemerged loud and clear in the late 1950s as part of the movement to end nuclear weapons testing. While it is quite true that key scientists like Linus Pauling were instrumental in initiating and continuing to pursue a nuclear weapons test ban treaty, the public exposure to the scientific debates and "scientific" threats from radioactive fallout left a significant measure of "antiscientism" in the active parts of the public. This was exacerbated, of course, by the role taken by the science-based defenders of weapons testing epitomized by physicist Edward Teller. Certainly the very popular film *Dr. Strangelove* (1964), which centered around the unsympathetic figure of the physicist-strategist who ultimately led the United States to unleashing the weapons of a massively destructive nuclear war, added a further negative image of scientists and their role in the Cold War world.[22] The antinuclear war movement, although including scientists among its early organizers and continuing active members, harbored a strong critique of science and its overly close liaison with the military. This movement, also largely middle class and involving younger people in increasing numbers, merged easily into the popular movement that sprang up in opposition to the war in Vietnam, especially when that war was expanded by President Lyndon Johnson in the mid-to-late 1960s. What at first came as praise to scholars and scientists for their designs and new strategies for "counter-insurgency" and new war fighting techniques, such as "the electronic battlefield," erupted finally on American campuses in 1968 and 1969 as a sharp sustained attack on the war, the political leadership judged responsible for it, and the scientists and their seemingly compromised position through their ties to the military and Department of Defense funding of their laboratories. More than one lab was attacked or threatened, others were picketed, and many universities were forced to rethink the nature of their relationships to defense agencies in specific and the federal government in general.[23] Along the way something else happened; the participatory politics, direct action campaigns, and civil disobedience protests initially nurtured by the Civil Rights movement flowed easily into the antiwar and then environmental movements. The new mood that came with it was opposition to authority and distrust of experts, scientific experts included. A call for relevance in knowledge and scholarship was accompanied by a desire to reintroduce ethics, values, and normative judgment into public policy and civil life. Radical social criticism was emboldened, alternative knowledge and practice sought, and dissenting intellectual activity praised.[24]

Within the broadly construed field of the historical and social study of science, a series of books, each achieving semipopular status, provide elements of the intellectual substrate, which laid challenge to the idea that science and scientific knowledge occupied a special or privileged position. Thomas S. Kuhn's *Structure of Scientific Revolutions* (1962) had the effect of undermining the idea that science, as distinct from other intellectual enterprises, produced a form of knowledge with special characteristics, perhaps enjoying a higher truth value. Kuhn's epistemologic challenge deprived science of special status and left the impression with many readers that "truth," if it had any meaning for science, would be defined by a given consensus or broad acceptance of an organizing paradigm;[25] revolutions in science occurred when the consensus shifted from one paradigm to another. There was, however, no necessary directionality and no obvious accumulation of knowledge along a single path. While Kuhn has objected to the idea that his work led to a relativistic approach to science, many readers took it in exactly that fashion.

Although Derek J. de S. Price's study *Big Science, Little Science* (1963) could be read at one level as a celebration of the new position science had achieved – i.e., "bigness" – at another level Price clearly lamented the loss of innocence and the special characteristics of a small, intimate activity.[26] Growth could almost seem disease-like and the control of the enterprise was becoming more distant from its practitioners. This, of course, was one of the critical points made two years earlier by President Eisenhower.

It might not seem that a study of *Giordano Bruno and the Hermetic Tradition* (1964), written by Renaissance scholar Frances Yates, would speak to the issue of the critique of science in the 1960s but that is one of its main points.[27] Yates ended her book on Giordano, that fascinating and clearly unorthodox student of the sciences, by lamenting the "positivist compromise" that marked the discourse of knowledge from value in the natural science. This compromise which Dame Yates dates to the rise of modern science in the late-Renaissance – early modern era would lie at the basis of many of science's future problems; the separation of the normative, the subjective and the magical from the "positive" knowledge of the science, she felt, would weaken the enterprise as a whole.

Daniel Greenberg's investigative journalist's report on *The Politics of Pure Science* (1967) is written in the best muckraking tradition; it exposed the scandals of the government/science relationship.[28] In some ways it serves as a case study of the abuses that Eisenhower seemed to predict. It was all the worse in Greenberg's eyes because it was scandal not among the engineers and technologists but rather in the academy and involving the "pure" scientists.

A number of other books could have been substituted for those mentioned above, but this collection serves as an example of the questioning of science widely spread through the literature that one could consider more internal to science studies. The misuse and abuse of science was certainly highlighted, but of greater importance for this study the status of the knowledge system itself was being undermined.

In chapter six of *One Dimensional Man* (1964), Herbert Marcuse, disciple of the Frankfurt School of political sociology and guru of the nascent "new Left" of 1960s intellectuals in the United States, advanced his critique of the "technological universe."[29] He saw in technological rationality a comprehensive and powerful mode of activity, which "shapes the entire universe of discourse and action, intellectual and material cultures." There were two parts to Marcuse's critique; the first, which had roots in earlier commentaries on the sciences, was that modern scientific rationality is inherently instrumentalist. Indeed, it was this very characteristic that was praised by others as the means by which science gained its putative objectivity. Marcuse's second criticism went beyond this to see instrumental rationality as providing the impetus behind "a specific technology, namely technology as a form of social control and domination." This leads Marcuse to his strongest critical statement in relation to science. "Science by virtue of its own method and concepts has projected and promoted a universe in which the domination of nature has remained linked to the domination of man."[30] That Marcuse saw the problem inherent in the very way science is known is made clear by his recommendation for change: alter the very form of scientific rationality itself; with this "science would arrive at essentially different concepts of nature and establish essentially different facts."[31] He was not requiring a mere reform in the manner in which scientific knowledge was used but rather saying that the very core of the scientific enterprise was flawed, that it needed to be reformed, and that the reforms would indeed so alter the practice of science that in a new science the very "facts" and "concepts" would be different. This represented a dramatic, radical departure from the traditional Marxist approach to the natural sciences where blame for the misshapen productions of the enterprise was placed on the deformation of science brought about by its practice under capitalistic and exploitative systems.

But Marcuse did not remain committed to this most radical critique of science (albeit segments of the younger new Left held on to it), responding in part at least to criticisms from some of his close intellectual associates. In his essay of clarification published in 1967 as "The Responsibility of Science," he partially retracted his criticisms of the core of science; instead, he reconstructed a contextualized version of the means to achieve a new practice for science. He made very clear his belief that, "There is no possibility of a reversal of scientific progress, no possibility of a return to the golden age for "qualitative" science. He did not want to be interpreted as seeking a revival of a Romantic philosophy of nature."[32] Instead, he linked the changes in science to changes in society.

The transformation of science is imaginable only in a transformed environment: a new science would require a new climate wherein new experiments and projects would be suggested to the intellect by new social needs. [These he set out directly]: Instead of the further conquest of nature, the restoration of nature; instead of the moon, the earth; instead of outer space, the creation of inner space.[33]

The structures of the science of the present, he claimed, were determined by the temporal and social context of conflict, war, ideological mobilization – the circumstances of the cold war and the contexts modern capitalism. He hoped instead that a new science would be guided by alternate goals – peace, happiness, and environmental beautification. Changes in the internal rationality of science would emerge only as these new goals established new priorities for the distribution of efforts and resources in the practice of science itself.[34] In this Marcuse seems to be responsive to Horkheimer's earlier claim that the link between scientific rationality and political domination came from the insistence that only one specific method of science was legitimate for the production of objective knowledge.[35]

Marcuse goes an important step further in demonstrating that under the proper conditions science and technology can play liberatory roles, that in a new social context technical reason itself "can become the technique of liberation."[36] By 1969, in *An Essay on Liberation*, Marcuse restates this evolved view of the relation between scientific-technical rationality and society in the following manner:

Is it still necessary to state that not technology, not technique, not the machine are the engines of repression, but the presence in them of the masters who determine their number, their lifespan, their power, their place in life, and the need for them? Is it still necessary to repeat that science and technology are the great vehicles of liberation, and that it is only their use and restriction in the repressive society which makes them into vehicles of domination.?[37]

This, of course, is almost a total reversion to the abuse of science and technology model easily agreed to by many contemporary critics. These later protestations never quite overcame the force of his earlier claims about the inherent flaws of scientific rationality and the belief that the techniques of domination lay therein. But Marcuse seemed to harbor relativist views and in his suggestion of sources for breaking the link between technique and domination he points to the possible resort to the cultural traditions of the non-Western world as a corrective to the destructive and repressive uses of modern technology.[38] There is a ring here of the voices who were almost simultaneously advocating the values of "appropriate technology" as a means of avoiding the pitfalls of Western technological development. In fact, Marcuse seemed to suggest that if societies of the Non-West succeeded, through their use of indigenous cultural values, in developing a nondestructive technology, it might well be reexported to the West as a worthwhile corrective.

If ambiguity remained for Marcuse and the Frankfurt legacy as to whether the dangers of science were inherent in the very system of knowledge itself or external to that system and shaped by the socio-political contexts of society, no such doubt existed for the "makers of the counterculture." These were the youthful "anti-intellectuals" whom Holton and other liberal defenders of science so feared. Theodore Roszak, the young historian, who took up their advocacy, opened his book *The Making of a Counter Culture* (1969) with an epigram drawn from William Blake.

Rise up O Young Men of the New Age! Set your foreheads against the ignorant hirelings! For we have Hirelings in the Camp, the Court, and the University, who would if they could forever depress Mental and prolong Corporeal War.[39]

Roszak's was open challenge to the intellectual establishment and admittedly "an outrageous pessimistic book." He took as the central target of his attack "The Technocracy" which he saw in the regime of experts and "think tanks" that reduced all humans in the words of Jacques Ellul to being "technical animals."[40] It is here that Roszak's link to the critique of science became clear. Technocracy, he opined, is "that society in which those who govern justify themselves by appeal to scientific forms of knowledge. And beyond the authority of science, there is no appeal."[41] (This of course is the "underside" of Ezrahi's explanation of science as the legitimator of authority in the democratic state; in this case science is cast as the court of final appeal.) Roszak took seriously the link between totalitarian control and science that had been advanced in Marcuse's early assessment but then largely abandoned (but not before the link had been adapted by many of those engaged in political and social dissent). The argument advanced by Roszak was that technocracy/science through a "regime of experts" helped to perfect the techniques of coercion, "... by exploiting our deep-seated commitment to the scientific world-view and by manipulating the securities and creature comforts of the industrial affluence which science has given us."[42]

For our discussion the major focus was the "counterculture's" attack on what it called the "Myth of Objective Consciousness," the title given to "Chapter Seven" of *The Making of a Counter Culture*.[43] Roszak supplemented this chapter with an Appendix, "Objectivity Unlimited," in which he set out illustrations of "the psychology of objective consciousness."[44] By juxtaposing objective consciousness with what he believes has been left out of science, "modes of nonintellective consciousness" (subjectivity?), he explicitly questions not only what science beholds but rather its very mode of viewing; it is a fundamental challenge to the *methods* of science themselves. The youthful dissenters of the counterculture, Roszak declares, turn from the idea of objective consciousness "as if from a place inhabited by the plague."[45] He calls up Thomas Kuhn to bolster his contentions about science, pointing to Kuhn's lack of conviction that scientific method is as purely rational or empirical as scientists like to claim.[46] The upshot of this argument is to de-privilege science and its way of knowing; Roszak's contention is "that objective consciousness is emphatically *not* some manner of definitive, transcultural development whose cogency derives from the fact that it is uniquely in touch with the truth."

Instead, he can conclude that science is "rather like a mythology, it is an arbitrary construct" which he locates in the context of a given society and its historical situation. And like any other mythology it is open to challenge by "cultural movements" that seek value and meaning from other sources. This is the course that the counterculture has chosen. For science the target is objective consciousness and its roots. The Enlightenment, "the entire episode of our cultural history, the great age of science and technology ... " stand "revealed in all its quaintly arbitrary, often absurd, and all too painfully unbalanced aspects."[47]

While obviously pleased to be able to recruit Michael Polanyi to his attack on objectivity, Roszak wants to go beyond Polanyi's doubt that objectivity actually exists even in the physical sciences.[48] Instead, he argues that independent of the epistemological status of objectivity, "it has become the commanding life style of our society: the one most authoritative way of regarding self, others and the whole of our enveloping reality."[49] Our personalities have become developed to feel and act as if objective observation was possible. This should be the focus of criticism. Experts should be rejected, for when it comes to what Roszak calls "the reality of our nonintellective (?) power," there are no experts and each person can be equally open to experience. In place of science, its epistemology and its psychology, he proposes that we open ourselves up to a "visionary imagination." The healer, the shaman, the magician should be revalued and the traditional skepticism born of science cast aside.[50]

The attitude and the outlook developed and nurtured within the counterculture profoundly rejected the "project of modernity." In this sense, Holton's fear of what he saw lurking alongside Mumford's critique of science was partially borne out by what the counterculture and its literary representatives said.

The youthful antagonists of science rejected the elites and the experts and insisted instead on participatory modes. By insisting on the unlikelihood of objective consciousness, and being suspicious of it, they replaced it with normative, subjective and nonintellective consciousness. Science was reduced in stature, removed from its position of power, influence, and authority; it was consciously de-privileged and stripped of its position as the ultimate source of understanding.

Lewis Mumford was not part of any "movement"; in fact, he was a profoundly nonpolitical individual. While he did participate in a number of anti-Vietnam war "teach-ins," he generally fled from the hall after his talk. Although he did take part in several antiwar marches, authorship was his chief mode of protest. He was one of the first American intellectuals to raise his voice against the war in Vietnam in 1965, and on a number of occasions clearly angered associates and friends for whom his vigorous attacks on the war and on President Lyndon Johnson seemed out of place.[51] But we also know that Mumford's own personal style left him uncomfortable with the youthful counterculture as it took up its own opposition to the war and antiestablishment actions. He was troubled by the personal messiness and the very impolite political activities and moods of the antiwar movement. Despite the type of fears expressed by Holton, Mumford never became a "cult figure" for the rebellious young (in this way distinct from Herbert Marcuse who did achieve such status). In a statement Mumford sent to a protest meeting at Berkeley, California, he praised his youthful audience. "You have awakened your country," but he went on to admonish them, "I am concerned with ... your reaction to violence." He further admonished the protesters, "No angry shouts; no ugly threats; no childish obscenities ... no mutilation of your minds by drugs."[52] It was the war, the war system, and the technology and science in it that incensed Mumford. This was part of a campaign he had opened just after the atomic bombs were dropped on Hiroshima and Nagasaki.

Mumford's first serious consideration of the question of technology came in his oft-cited book *Technics and Civilization* (1934).[53] Written during the troubled days of the Great Depression, when calls were being issued for a moratorium on technology, Mumford had already introduced a note of worry about technological developments, especially in the present "neo-technic" era. But science, he believed was different and had in it a potential for restraint, even while it was providing the new knowledge of technological and industrial advance. Science, even in its earlier history when it was "remote and by nature esoteric, it was not snobbish. Socialized in method, international in scope, impersonal in animus, performing some of its most hazardous and fruitful feats of thought by reason of its very divorce from immediate responsibility . . . "[54] He was almost euphoric as he looked back at the period of the Scientific Revolution, what it had wrought, and the attitudes it had developed. The atomic bomb and its use changed all that, however, and evoked from Mumford a profound distrust for the bomb's scientific builders. His very first article on the subject, "Gentlemen You are Mad" (1946) indicated the degree of disillusion that had come to mark Mumford's attitude toward the scientists themselves.[55]

By the 1960s, it was not only scientists as morally frail individuals to whom Mumford pointed but the core of science itself. He began to reject the outcome of the Scientific Revolution, noting that the problem was not one of omission alone but actually of commission. By rejecting subjectivity and by praising objectivity alone, science, he felt, had dismissed humans themselves. Mumford was troubled by the depersonalized worldview that had accompanied the rise and increasing importance of science. He seemed fearful of the overorganization of modern life. He decried the "well-organized system under centralized direction, which achieves the highest degree of mechanical efficiency when those who work it have no mind or purpose of their own."[56] Several of Mumford's deeper held beliefs come together here. His concern for both the personal and the subjective appeared to be a victim of the overrationality and the depth of commitment to the "objective" that Mumford links to successful modern science. While he obviously had a life-long fascination with machines, he also had an equally long commitment to an organic as opposed to a mechanical worldview.

Another theme, present in his earliest writings on technics, emerged in the 1960s in time for it to join a profound new social and popular awareness of the democratic and the participatory.

My thesis, to put it bluntly, is that from late neolithic times in the Near East, right down to our own day, two technologies have recurrently existed side by side: one authoritarian, the other democratic, the first system-centered, immensely powerful, but inherently unstable, the other man-centered, relatively weak but resourceful and durable.[57]

The challenge that he saw emerging in the mid-1960s was that democratic technics was being overwhelmed and "every residual autonomy . . . wiped out" or at best totally marginalized. This lament, which he clearly shared with Dwight Eisenhower, he now rooted in his reinterpretation of the history of science and technology. The envisaged victory of authoritarian and centralized technics relied upon the scientific discoveries and inventions so often praised for the advancements they brought but also upon the creation of "human machines," the bureaucracy, the military army, and the work army.[58] While the growth of experimental science had been seen by optimistic prophets like August Comte and Herbert Spencer as the guarantor of peaceful, productive and democratic societies, today the outcome is almost a reverse. Mumford's charge is blunt:

with the knitting together of scientific ideology, itself liberated from theological restrictions or humanistic purposes, authoritarian technics found an instrument at hand that has now given it absolute command of physical energies of cosmic dimension.[59]

The rocket builders, atomic scientists, and computer designers are compared to the pyramid builders of ages past; "boasting through their science of their increasing omnipotence . . . omniscience," they are driven by obsessive, compulsive irrationalities, careless of the ultimate costs of life. They are willing to move toward the "absolute instruments of destruction" and even the "mutilation or extermination of the human race." So angry was Mumford at this new authoritarian technics that he proclaimed that by comparison, "Even Ashurbanipal and Genghis Khan performed their gory operations under normal human limits."[60]

Like the young radicals who would later develop their own vocabulary of criticism, Mumford in the early and mid-1960s focused over and over again on the new *system* in which authority now resided – no longer in a monarch or a totalitarian dictator, but in the *system*. Even the "sacred priesthood of science who alone have access to the secret knowledge by means of which total control is now swiftly being perfected" have become enmeshed in the very organization-system that they created. Mumford introduces his term "Pentagon of Power" in this 1964 essay on the threats of authoritarian technics; it presents a "systems-centered collective" depersonalized with no visible person/deity who commands.[61] The link to science is direct. Not that some new discoveries or inventions caused the threat to democracy; but rather since the time of Francis Bacon and Galileo when the new science set out its methods and goals, the great transformations that have occurred have been guided by a system in which human personality and historic process have been replaced by abstract intelligence, a system in which control over nature and ultimately humans became the "chief purpose of existence." What troubles and perplexes Mumford is that we have "surrendered so easily" to the controllers of the authoritarian technic. He views the whole arrangement as a "magnificent bribe." Material advantages of all sorts will become available, far beyond the elites who previously could command them, in this new "democratic-authoritarian social contract"; but once accepting the terms of the contract "no further choice remains." As Mumford poses the trade-off, "Is it really humanly profitable to give up the possibility of living a few years at Walden Pond, so to say, for the privilege of spending a lifetime in Walden Two?"[62]

The new age that Mumford feared was on the way was an age of overorganization dominated by elites and experts who, having signed the contract, were noticeable for their monopolization of technical knowledge and a regimented work organization and life organization. Thus is created the loyal bureaucrat, the Robot or, as Mumford acknowledges, the "Organization Man" described so clearly by T. H. White in his classic study of large-scale corporate enterprise in the United States.[63] As scientists adopted the way of the Organization Man they would become little more than a "depersonalized servo-mechanism in the megamachine." They would become capable of any task, able to follow any command, and make no judgments of value about their work or express moral misgiving. It was on those terms Mumford claimed that "Adolph Eichmann, the obedient exterminator, who carried out Hitler's policy and Himmler's orders with unswerving fidelity, should be hailed as the 'Hero of our Time'."[64] Scientists entering this state were becoming a menace to global survival. As Mumford put it, there was an "Eichmann in every missile center ready to obey orders, even if horrific."[65] "There are now," Mumford lamented, "countless Eichmanns in administrative offices, in business corporations, in universities, in laboratories, in the armed forces" ready to obediently carry out "any officially sanctioned fantasy," independent of its horror or harm.[66]

With so deeply a pessimistic view of his contemporary world and his fellow well-trained, educated, and highly skilled human beings, was there any way to overcome the crises Mumford had so forcefully delineated? Readers of Mumford's *Pentagon of Power* must have been as surprised as was his reviewer, Lehmann-Haupt, when in the final pages of his angry critique he abandons his pessimism and points to optimistic signs. "Suddenly, two thirds of the way through his book, he is optimistic about things as they are."[67] But for Lehmann-Haupt, Mumford's "sugar-coated pill" sounds unconvincing, "a flimsy afterthought to his overwhelmingly gloomy portrayal of 'the megatechnical wasteland.'" Mumford's proposal seems remarkably simple, even naive; it carries in it something of the same spirit as governed much of the politics of the counterculture. Mumford notes that those who remained in the system had been bribed with privilege, pleasure, and affluence; but he added, with prophetic tone, "for those of us who have thrown off the myth of the machine, the next move is ours; for the gates of the technocratic prison will open automatically as soon as we choose to walk out."[68] Withdrawal and conversion seemed to be Mumford's answer; not Marx but Thoreau. For Mumford the sources of the new worldview which may, during the next generation, succeed in bringing under control the destructive forces of science are found in "the new organic model of ecological association and self-organization (autonomy and teleonomy)" which were brought together (unknowingly) by Charles Darwin in his theory of evolution.[69] In this Mumford returns to type, an organic world, embodying the properties of organisms/organic systems – qualitative richness, amplitude, spaciousness, self-regulation, self-correction, self-propulsion. Balance, wholeness, completeness, continual interplay between the inner and the outer, the subjective and the objective – the whole catalogue of organismic goodness is outlined, and in some detail.[70]

Mumford provides another interesting clue to his ideas of what might make a difference in the outlook of people toward the authoritarian technology of the late twentieth century. He recounts the story of the electric power failure that blacked-out much of the northeast United States in November 1965, and most important for him, the people's response.

Suddenly as in E. M. Forster's fable, *The Machine Stops*, millions of people caught without either power or light, immobilized in railroad trains, subways, sky scraper elevators, moved spontaneously into action, without waiting for the system to recover or for orders to come from above.[71]

You can almost hear the enthusiasm in his voice as he cites the words from the *New Yorker*, "while the city of bricks and mortar was dead, the people were more alive than ever." The very "Foreword" to his book is repeated once again, in place of "Mechanization takes Command," the leitmotif of the past few centuries, Mumford approvingly quotes the worlds of astronaut John Glenn, "Let Man Take Over."[72]

The image of change that lies behind Mumford's last-moment optimism is human spontaneity, not rationally planned and bureaucratically enacted activity. "For its effective salvation mankind will need to undergo something like a spontaneous religious conversion ..."[73] The mechanical worldview will be replaced by an organic worldview, human personality will be given precedence over machines and computers. The support he calls upon from history is a reminder that Christianity replaced the "classic power complex of Imperial Rome." Other such changes he contends have often occurred throughout human history. But of one thing he is certain, "If mankind is to escape its programmed self-destruction, the God who saves us will not descend from the machine: he will rise up again in the human soul."[74]

A religious-style conversion, not mass political action, will bring the new world Mumford sought; an organic and ecological worldview, not socio-political analysis of the structure of society, will provide the guide and substance of the new worldview. Mumford was at one level profoundly in tune with the critiques of modernity, technics, and science that came alive in the 1960s. In a peculiar way, his solutions matched only those held by the reformers at the margins of the protest movements of that era. His views – personalist, religious, organic – were not those of the new Left or the rebellious radicals; they did, however, embody some of the same return to "selfness" of the counter-culture and the organismic outlook of the nascent environmentalists; his disdain for the bigness of large-scale science and technology and its antidemocratic tendencies became the core of criticism embodied in *Small is Beautiful* (the text by E. F. Schumacher) and the movement for the development of "appropriate technologies."[75]

Did the pessimistic assessments and critiques of the decade of the "sixties" have any real-time or lasting effects? The military megamachine with its increasingly destructive and desperate weapons rolled ahead unimpeded by criticism and campaign, continuing to enlist the "best and the brightest" to its ranks; it slowed down only in response to the collapsing power of the Soviet Union and the end of the Cold War. The desire to eliminate and control nature seems unabated, from genetic engineering to artificial intelligence. The one area where at least recognition of the problem has become apparent, concerns ecology and environment. In this domain, both detailed examination and some promise of action, regulation, amelioration can be found at the national level in many countries and in preliminary form at the international level. Public attitudes have undergone considerable change, not yet matched, however, by public action. While an institutionalization of interest in the social responsibility of science and scientists has become manifest in new programs, publications, and periodic pronouncements, it is an activity marginal to the day-to-day actions of institutionalized science and technology. Science still maintains a privileged position even if there has been some increase in the scrutiny of ethical and policy issues. Science and technology seem no more "human" in scale now than before. Eisenhower, Carson, Mumford, Marcuse, and Roszak would hardly be expected to alter their pessimistic assessments on the bases of current thought and practice.

Notes

1. Lewis Mumford, *The Pentagon of Power*, Vol. 2 of *The Myth of the Machine* (New York: Harcourt, Brace, Jovanovich, 1970).
2. *Ibid.*, "Preface"; Lewis Mumford, *Technics and Civilization* (New York: Harcourt, Brace, 1934).
3. Lewis Coser, review of *Pentagon of Power*, *Contemporary Sociology* 1 (1972), 38–39; see Lewis Mumford, "Call Me Jonah," *My Works and Days* (New York: Harcourt, Brace, Jovanovich, 1979), pp. 527–531.
4. See Everett Mendelsohn, "Prophet of Our Discontent: Lewis Mumford Confronts the Bomb," in Thomas P. Hughes and Agatha C. Hughes (eds.), *Lewis Mumford: Public Intellectual* (New

York: Oxford University Press, 1970), pp. 343–360.

5. Mumford recounts these comments in a letter to Hilda L. Lindley, October 17, 1970, cited in Hughes, *Lewis Mumford*, p. 7; see also Donald L. Miller, *Lewis Mumford: A Life* (Pittsburgh: University of Pittsburgh Press, 1989), p. 534.

6. Cited in Miller, *Mumford*, p. 534, from "Random Notes," September 16, 1970.

7. Cited in Miller, *Mumford*, p. 535, from Mumford's response to Holton's review.

8. Yaron Ezrahi, *The Descent of Icarus: Science and the Transformation of Contemporary Democracy* (Cambridge, Mass.: Harvard University Press, 1990).

9. See Adam Goldgeier's review in *Tikkun* (July/August 1992), 59–62. See also Everett Mendelsohn's early essay, "Should Science Survive Its Success," in Robert S. Cohen, et al. (eds.), *For Dirk Struik* (Dordrecht: Reidel, 1974), pp. 373–389.

10. Cited in William Leiss, "The Social Consequences of Technological Progress: Critical Comments on Recent Theories," *Canadian Public Administration* 13 (1970), 249, from Emanel G. Mesthene, "Technology and Wisdom," *Technology and Social Change* (Indianapolis: Bobbs-Merrill, 1967), p. 59.

11. Jonathan Allen (ed.), *March 4: Scientists, Students and Society* (Cambridge, Mass.: MIT Press, 1970), "Union of Concerned Scientists, Faculty Statement," pp. xxii–xxiii.

12. Dwight D. Eisenhower, "Farewell Radio and Television Address to the American People, January 17, 1961," *Public Papers of the President of the United States, Dwight D. Eisenhower, 1960–61* (Washington DC: Government Printing Office, 1961), pp. 1035–1040.

13. C. Wright Mills, *The Power Elite* (New York: Oxford University Press, 1956).

14. Rachel Carson, *Silent Spring* (Boston: Houghton, Mifflin, 1962), facing the title page.

15. See Samuel D. Hays, *Beauty, Health, and Permanence: Environmental Politics in the United States, 1955–1995* (Cambridge: Cambridge University Press, 1987).

16. Carson, *Silent Spring*, p. 261.

17. Cited by her publisher-biographer, Paul Brooks, *The House of Life: Rachel Carson at Work* (Boston: Houghton, Mifflin, 1972), p. 128.

18. CBS Special, "The Silent Spring of Rachel Carson."

19. See Donald Fleming, "Roots of the New Conservation Movement," *Perspectives in American History* 6 (1972), 7–91.

20. Gregg Herken, *The Winning Weapon: The Atomic Bomb in the Cold War, 1945–1950* (New York: Alfred Knopf, 1980).

21. See Mendelsohn, "Prophet of Our Discontent."

22. See Spencer Weart, *Nuclear Fear, A History of Images* (Cambridge, Mass.: Harvard University Press, 1988).

23. Tom Bates, *Rads: The 1970 Bombing of the Army Math Research Center at the University of Wisconsin and its Aftermath* (New York: Harper/Collins, 1992).

24. See Theodore Roszak (ed.), *The Dissenting Academy* (New York: Pantheon Books, 1968).

25. Thomas S. Kuhn, *The Structure of Scientific Revolutions* (Chicago: University of Chicago Press, 1962).

26. Derek J. Price, *Little Science, Big Science* (New York: Columbia University Press, 1963).

27. Frances A. Yates, *Giordano Bruno and the Hermetic Tradition* (Chicago: University of Chicago Press, 1964).

28. Daniel S. Greenberg, *The Politics of Pure Science* (New York: New America Library, 1967).

29. Herbert Marcuse, *One Dimensional Man* (Boston: Beacon Press, 1964); see also William Leiss, "Technological Rationality: Marcuse and His Critics," *Philosophy and Sociology of Science* 2 (1972), 31–42.

30. Marcuse, *One Dimensional Man*, p. 166; also cited in Leiss, "Technological Rationality," p. 37.

31. Marcuse, *One Dimensional Man*, p. 167; Leiss, "Technological Rationality," p. 38.
32. *Ibid.*, p. 38; cited from Marcuse, "The Responsibility of Science," in L. Krieger and F. Stern (eds.), *The Responsibility of Power* (New York: Doubleday, 1967), pp. 442–443.
33. *Ibid.*
34. *Ibid.*
35. Leiss points to Max Horkheimer, "Zum Problem der Wahrheit," in A. Schmidt (ed.), *Kritische Theorie* (Frankfurt, 1968), Vol. 1, p. 259.
36. Leiss, "Technological Rationality," p. 39, cites the remark from Marcuse, "Industrialization and Capitalism in the Work of Max Weber," in Herbert Marcuse, *Negations* (Boston: Beacon Press, 1968), pp. 222–223.
37. Herbert Marcuse, *An Essay on Liberation* (Boston: Beacon Press, 1969), p. 12.
38. Leiss, "Technological Rationality," p. 40 and Herbert Marcuse, *Soviet Marxism* (London: Routledge [1958], rpt. 1969), Preface, p. xvi.
39. Theodore Roszak, *The Making of a Counter Culture: Reflections on the Technocratic Society and Its Youthful Opposition* (New York: Doubleday, 1909), p. ix.
40. *Ibid.*, pp. 5–6.
41. *Ibid.*, pp. 7–8.
42. *Ibid.*, p. 9.
43. *Ibid.*, pp. 205–238.
44. *Ibid.*, pp. 269–289.
45. *Ibid.*, p. 215.
46. *Ibid.*, pp. 213–214.
47. *Ibid.*, p. 217.
48. Roszak is referring to Michael Polanyi, *Personal Knowledge: Towards a Post-Critical Philosophy* (Chicago: University of Chicago Press, 1959).
49. Roszak, *Making of a Counter Culture*, p. 216.
50. *Ibid.*, pp. 236ff.
51. Miller, *Mumford*, pp. 513ff.
52. Cited in Miller, *Mumford*, from a statement prepared for, but not actually read at, the meeting, p. 517.
53. Lewis Mumford, *Technics and Civilization* (New York: Harcourt, Brace, 1934).
54. *Ibid.*, p. 408.
55. Lewis Mumford, "Gentlemen You Are Mad!" *Saturday Review of Literature* 2 (March 1946), 5–6. See also Mendelsohn, "Prophet of Our Discontent."
56. Lewis Mumford, "Authoritarian and Democratic Technics," *Technology and Culture* 5 (1964), 2.
57. *Ibid.*
58. *Ibid.*, p. 3.
59. *Ibid.*, p. 5.
60. *Ibid.*
61. *Ibid.*, p. 6.
62. *Ibid.*, p. 7.
63. Mumford, *Pentagon*, p. 278.
64. *Ibid.*, pp. 278–279.
65. Cited by Miller, *Mumford*, p. 539.
66. Mumford, *Pentagon*, p. 435.
67. Lehmann-Haupt, review *New York Times*, November 9, 1970.
68. Mumford, *Pentagon*, p. 435.
69. *Ibid.*, pp. 393ff.

70. E.g., *Ibid.*, p. 435.
71. *Ibid.*, p. 412.
72. *Ibid.*, and p. vii.
73. *Ibid.*, p. 413.
74. *Ibid.*
75. E. F. Schumacher, *Small is Beautiful: A Study of Economics as if People Mattered* (London: Blout & Bryss, 1973), and George McRobie, *Small is Possible* (New York: Harper and Row, 1981). A factual account about who is doing what and where to put into practice the ideas expressed in E. F. Schumacher's *Small is Beautiful*.

THE CULTURAL CONTRADICTIONS OF HIGH TECH: OR THE MANY IRONIES OF CONTEMPORARY TECHNOLOGICAL OPTIMISM

HOWARD P. SEGAL
University of Maine at Orono

When I published my study of *Technological Utopianism in American Culture* in 1985, I thought that the historical phenomenon I detailed there – an uncritical faith in technology's ability to solve all problems – would soon be a relic of a more hopeful, if more naive, era. I believed that increasing numbers of Americans, following the lead of Europeans and others, had begun to seek healthy limits on unadulterated technological advance and were starting to express greater concern for nontechnological realms crying out for equivalent attention, enthusiasm, and funding. I sensed movement toward what I called a "technological plateau." I still do, but my optimism, never overwhelming, has since been tempered further.

As I detailed in my book, the commonly held conception of America as a potential utopia, and a utopia to be brought about by technological progress, is an old and familiar one and has many European roots. It long predates the actual technological wonders that for so many in the United States and abroad promised to transform utopian dreams into reality. And it is a conception apparently still shared by many Americans (and non-Americans) even as doubts over both America's and technology's future continue to grow. For all the now familiar criticism of technology in recent decades generated by such developments as the proliferation of atomic weapons, the pollution of the natural environment, and the malfunction of nuclear power plants, there remains considerable faith in American technology's ability to solve problems and to improve society.[1] This faith was evidenced most recently by the national outpouring of zeal for the computerized weapons systems that allegedly proved so decisive in the Persian Gulf War. Overnight, it seemed, any doubts about the prowess of military technology evaporated amid numerous examples of its "high tech" achievements. Suddenly the supposedly dormant spirit of technological utopianism appeared almost everywhere. The fact that, in retrospect, the capabilities of those weapons systems were greatly exaggerated has not yet (wholly) dampened the spirit of spring 1991. Hence there is some basis for the otherwise questionable recent advertising claim of a high tech company that technology is still "the romance of our age." (High tech's need for historical legitimacy in its advertising is explored in section II.)

All of this is not, however, to dismiss technology's contemporary critics, for I count myself among them. Nor, for that matter, is it to endorse what I and others, paraphrasing historian Herbert Butterfield, like to call the overly optimistic "Whig Theory" of the history of technology.[2] Rather, it is to suggest the extraordinary complexity of a fundamental problem: the relationship, or lack thereof, between technological progress and social progress. It is also to suggest the continuing utility of placing current concerns for this relationship in a historical context.

In this regard, I wish to examine that recent resurgence of technological utopianism associated with the Persian Gulf War and to suggest both that it is not limited to high tech weapons systems and that it predates that conflict. Rather, it is reflective of high tech in general, broadly defined as computers, satellite communications, robotics, space travel, genetic engineering, and other post-World War II technologies. Certainly other, less glitzy technologies have emerged since 1945 – in the chemical industry, for example. I exclude them not because they are less significant but because commonly accepted notions of high tech exclude them. More precisely, high tech is usually contrasted with traditional dirty, large-scale manufacturing and power facilities through its greater cleanliness and efficiency and its smaller size as well as through its allegedly "paperless" communications systems. High tech, as used here and elsewhere, fits neatly within the framework of a contemporary postindustrial society, given the high tech industries' foundation on information collection, analysis, and distribution.

Admittedly, "high tech" itself is a relative term applicable to earlier periods, not least of which is the prior industrial revolution of large-scale factories. Just as technology has been present to some degree in every society, so numerous prior eras have had their own "cutting edge" technologies. Still, contemporary high tech has its own particular identity and self-image and can be used without apology.

So defined, contemporary high tech has been and remains a notable exception to the perception of technology in much of the modern world as having gradually shifted from a social solution to a social question. High tech, by contrast, embodies and promotes otherwise largely discarded beliefs in progress per se and in the causal connection between technological progress and social progress. "Star Wars" is perhaps the foremost example of this persistent strain of technological utopianism associated with high tech, but it is hardly the only one, as the Persian Gulf War weapons systems example makes clear. High tech, in fact, appears not only as optimistic about the future but also more indifferent toward and, in other contexts, more manipulative of the past than earlier technologies have been. Indeed, high tech is eager to proclaim itself the supreme technological revolution while enjoying an unprecedented ability to articulate and spread its message, thanks to the very communications and transportation systems it exalts.

But this unqualified faith in the future, unlike that of earlier technological utopians, is only a facade. Behind it is a largely unacknowledged ambivalence about the future and, equally important, a sometimes desperate desire to connect to the past for intellectual legitimacy (even when this means manipulation of, or seeming contempt for, the past). In this context, I will examine four leading ways in which high tech promotes its products and its ideology: prophecies, advertising, world's fairs/theme parks, and the technological literacy crusade. However diverse these phenomena, they are linked by a common vision of technological utopia or at least of high tech's notion of the "good society."

I

High tech has spawned a new generation of technological utopians whose principal allegiance is not to the public sector, unlike earlier such visionaries, but to the private; whose favored institution is not big government but the big corporation; and whose principal motivation is not serious social change but personal gain – or prophecy for profits' sake. Once the province of generally well-intentioned if often fumbling amateurs, technological forecasting has become a business, an increasingly big business, populated with "professionals." Their clients are generally major corporations plus those government agencies and educational institutions who can afford their enormous consulting fees. These forecasters' success is a tribute to the spirit of free enterprise they espouse, and it is hardly surprising that their overall message is resolutely optimistic even when things seem to be going in the wrong direction. To the extent that they are considered, most national and international problems, these high tech prophets insist, will be solved through ever more and better technology and through the sheer determination to use it decisively.

The most popular high tech prophets, though hardly the only ones, are Alvin Toffler and John Naisbitt. It is a measure of high tech's underlying ambivalence about the future that their respective messages, however different, are so passionately embraced as gospel by business and professional people, who need the reassurance these prophecies amply provide. The conventional argument that upbeat and readable books about the future invariably outsell gloomy and jargon-filled ones only partly accounts for Toffler's and Naisbitt's extraordinary successes. Not only do doomsday prophecies often also become best sellers, but Toffler's inventive vocabulary is as much a barrier to readability as it is a means to grasping his message. (Naisbitt writes flat, colorless prose, broken up repeatedly by topic sentences with bold letters.) Rather, high tech wants to be told that its products and services are bringing the world closer to some kind of utopia.

Like most earlier technological utopians, Toffler and Naisbitt largely extrapolate from today to tomorrow while showing painfully limited interest in the past. Yet, as with their predecessors, it is the unacknowledged past that invariably provides the actual basis of these contemporary prophecies, whether as the extrapolation source or as the supposed contrast with the great age ahead. Try as they may, they cannot escape history's grasp.

Toffler has written and edited several books about the future, the most popular of which are *Future Shock* (1970), *The Third Wave* (1980), and *Powershift* (1990), each neatly appearing at the start of a new decade. *Future Shock* was published five years after Toffler coined the term while writing an article. It was the first futurist best-seller.[3] In those five years he visited people and places concerned with change and coping behavior. The term "future shock" is variously defined in his book, but in general it means the cumulative effects of the acceleration of change on individuals and societies and the limitations on the amount of change that ordinary persons can absorb in a short time. It is a form of culture shock applied more to time than to space. According to Toffler, an unending number of changes in everyday life – enormous technological advances, information overload, immense diversity of choices and decisions, temporariness of human relationships, intensive urbanization, instant communications – have created an unprecedented "collision with the future." Future shock is the dizziness brought on by "the premature arrival of the future."[4] How the future can arrive ahead of schedule is beyond me and is surely a logical contradiction, but the more interesting point is the consequent separation of the future from the past and in turn the irrelevance of history to present and future alike.

At first glance, Toffler's description of the contemporary world, written in the breathless prose of *Time* magazine in its heyday (he once worked at *Fortune*), is rather depressing to contemplate. Toffler has taken the concept of "cultural lag" identified with sociologist William Ogburn (whom he characteristically simplifies and distorts) to an absurd degree. Ogburn suggested that disequilibrium results from the slowness with which nonmaterial culture adapts itself to technological advances. Neither personally nor organizationally, Toffler contends, can existing means of adaptability to change save us. Rather, we need new ways of looking at and responding to the future and, not surprisingly, new visionaries to guide us to the better world ahead. Toffler, of course, is happily available to save the world from "massive adaptational breakdown."[5]

Among other things, Toffler recommends that individuals turn off sensory stimuli and maintain personal stability zones, or patterns of relative constancy. In specific terms, this means, for example, turning on air-conditioners to lower street noise, seeking silence on deserted beaches, traveling less, keeping clothes and cars for additional years, and reorganizing companies, churches, and community groups less often. For societies he suggests educational reforms and the control of technology. Schools must cease preparing their students for the past, for the world of the nineteenth-century industrial revolution, and must instead literally teach the future. Information that will be useless for the future should simply be forgotten, and that presumably includes history. Significantly, Toffler rarely mentions and barely criticizes advertising as a major cause of future shock, with its endless creation of artificial wants. Advertising, perhaps not coincidentally, is the other source of his writing style.

Lest I misrepresent his position on technology, Toffler here, as in later writings, is no technological pessimist. He despises the so-called antitechnologists allegedly epitomized by Jacques Ellul, Erich Fromm, Herbert Marcuse, and Lewis Mumford. Modern technology, he argues, is the great "engine of change" in societies and the foremost guarantor of much of our freedom, particularly our freedom of choice and of diversity. "It is only primitive technology," he claims, "that imposes standardization." No technological plateau for him. Similarly, Toffler rejects as "myth" the vision of future "man as a helpless cog in some vast organizational machine." Instead, bureaucracy will vanish, replaced by "ad-hocracy" – short-term, professional, problem-solving task forces.[6]

The fact that untold individuals and societies throughout history have, despite enormous challenges and costs, adapted to changes as profound and as accelerated in their own day as those Toffler details for the 1970s naturally is of no interest to him. For then he would have precious little originality save perhaps regarding specific recommended adaptations to high tech. And the equally important fact that all technological developments, whether beneficial or harmful, are the products of human actions and decisions and not autonomous forces, is likewise something Toffler needs to ignore to sell his argument. Moreover, his predicted "death of permanence" hardly means the end of history and might be turned around to justify accelerated historical study just to keep track of so many changes and to contribute to the "soft landing" on the future he seeks.[7]

At the outset of *Future Shock*, Toffler makes a brief, carefully low-key disclaimer that can be applied to his later writings as well.

...I have taken the liberty of speaking firmly, without hesitation, trusting that the intelligent reader will understand the stylistic problem. The word "will" should always be read as though it were preceded by "probably" or "in my opinion." Similarly, all dates applied to future events need to be taken with a grain of judgment.[8]

This is hardly a mere "stylistic" issue, but it is certainly a problem. Like so many other "scientific" futurists, Toffler thereby qualifies his predictions, not only to avoid being wrong but also to concede his apparent ambivalence about the future. However plentiful his statistics, examples, and jargon, his prophecies are anything but guaranteed. Yet he wants his readers and other audiences to go along with the traditional American positive thinking that helps account for his works' popularity (and that places them in proper historical context). Positive thinking, in fact, gives a wonderful advantage to those business and professional people in a position to alter the world; it enables them to substitute willpower and self-improvement for the fundamental economic, social, political, and other changes that one might innocently assume essential to avert lasting future shock.

True, Toffler advocates continuing popular plebiscites on the future ("social future assemblies") and the elimination of traditional technological planning – not that technological planning, much less the social planning that should accompany it, has a long or popular history in the United States. Like Naisbitt, Toffler places unlimited faith in the capacity of mass communications, including television and computers, to liberate individuals and to invigorate democracy. Such faith was as clearly misplaced in 1970 as it is today. Toffler's failure to discuss political power in America and its future distribution is as revealing as anything else of his actual commitment in *Future Shock* to the political status quo. The "anticipatory democracy" he forecasts remains a pipe dream.[9]

Like all good futurists, Toffler actually deals with the present and the past as much as the future. "Future shock," to the extent it exists, is really "present shock." Paradoxically, so much contemporary upheaval should logically restrain, not prompt, serious speculation about tomorrow, lest further change undermine the prescriptions offered at the present moment. This is or should be the price to be paid for the unprecedented acceleration Toffler insists we accept as reality. But he chooses not to pay it.

In *Future Shock*, Toffler does allow for one small use of history: as respite from the present and the future. He argues on behalf of "living museums" where people could, in effect, take a holiday from change. For example, "children from the outside world might spend a few months in a simulated feudal village, living and actually working as children did centuries ago."[10] That "living history museums" already exist in the United States and elsewhere is something Toffler apparently is unaware of. The more telling point is his indifference to anything such museums might teach us, including just how fast or slow life really was in earlier times; Toffler obviously assumes it was profoundly slow. Once again, he cannot come to grips with the past even when, as here, he attempts to manipulate it for his own purposes.

At first glance, *The Third Wave* is more historically grounded, treating as it does the whole of history. The First Wave began about 8000 B.C., when roving bands of hunters, having learned to till the soil, settled in villages. It lasted until the Industrial Revolution of about 1650–1750 A.D. The Second Wave, which gave us the industrialized, standardized society that most of us grew up in, is itself giving way to the postindustrial Third Wave, which could last forever if only people listened to sages like Toffler. Among the Third Wave's characteristics are an increased emphasis on leisure, the decline of the nuclear family, the replacement of outdated political institutions and processes, and the growing use of genetic engineering. As in *Future Shock*, Toffler views technology as deterministic and for the most part happily so. Computers, word processors, and microprocessors are the 1980s' "engines of change." In fact, the Third Wave, even more than future shock, is beyond human control. "In a great historical confluence," Toffler contends, "many raging rivers of change are running together to form an oceanic Third Wave of change that is gaining momentum with every passing hour."[11] Where *Future Shock* emphasized the cost of rapid change, *The Third Wave* emphasizes the cost of not changing rapidly enough.

True, the Third Wave will utilize technology to create a new society combining some aspects of both of its predecessors. This includes the ability to work at home or nearby in high tech "electronic cottages" reminiscent of the First Wave; decentralized and "demassified" customized production derived from Second Wave centralized mass production and the restoration of producers and consumers, divorced in the Second Wave, into "prosumers" recalling the First Wave.[12] But the Third Wave will not celebrate any legacies of its predecessors but will instead wash them out into oblivion. The real possibility that new technologies might reinforce rather than subvert the status quo never occurs to Toffler. Meanwhile the transitions to the Third Wave will generate tensions and, if we are foolish, conflict and even self-destruction. Those with vested interests in the Second Wave will fight fiercely against the inevitable. Yet the eventual outcome is, of course, positive, if mankind changes its ways.

To his credit, Toffler here deals with anticipated changes in political power and process more seriously than in *Future Shock*. In the concluding chapter of *The Third Wave* he outlines a new form of decentralized participatory democracy appropriate for a "demassified" age. Majority rule by mass political parties is to give way to temporary "modular parties" reflecting shifting pluralities of various interest groups. Elected representatives may even be replaced or complemented by ordinary citizens using high tech communications systems to learn about issues and then to vote on them. Toffler seems genuinely concerned about the breakup of Second Wave governments and institutions possibly leading to authoritarianism. Hence the need, more urgent than in 1970, for "anticipatory democracy" to save the day.[13] Once again, however, his vision lacks a critique of the fundamental obstacles to genuine participatory democracy in the future, not least of which are multinational corporations.

"As Third Wave civilization matures," Toffler predicts, "we shall create not a utopian man or woman who towers over the people of the past, not a superhuman race of Goethes and Aristotles (or Genghis Khans or Hitlers) but merely, and proudly, one hopes, a race – and a civilization – that deserves to be called human." Yet Toffler believes that the Third Wave will evolve into a "practopia," or practical utopia.[14] Significantly, *The Third Wave* hedges its predictions much more than *Future Shock*, and Toffler is sometimes at pains to let his optimism overcome the pessimistic portrait he paints of an industrial civilization otherwise heading for chaos and collapse. As even Edward Cornish, president of the World Future Society and a Toffler admirer, conceded, he still wondered "just where Al spells out exactly what the Third Wave is," so deliberately qualified and vague is the content.[15] Once more Toffler's ambivalence about the future comes through.

For this reason Toffler needs to have the otherwise neatly separated Second and Third Waves overlap until the latter someday triumphs. Whatever uncertainty and confusion currently exists can be blamed on this temporary overlap. Any serious student of history would recognize the simultaneity yet today of agriculture and industrialization amid postindustrialization, and the likelihood of continuities among them for the foreseeable future. For Toffler, however, the Second Wave must finally give way entirely to the third, just as the first allegedly gave way to the second. Historical discontinuity is crucial for his argument, lest the future not roll in as he repeatedly insists it will.

At the beginning of *The Third Wave*, Toffler writes that "beneath the clatter and jangle of seemingly senseless events there lies a startling and potentially hopeful pattern. ... the human story, far from ending, has only just begun."[16] Notice the further forced separation of, in effect, all of history from the future, as if "the human story" had no past but only a (new) beginning; or, going further, as if no one had ever transformed the "seemingly senseless events" of earlier times into a coherent pattern. Revealingly, in *Previews and Premises: An Interview with the Author of "Future Shock" and "The Third Wave"* (1983), Toffler brands persons who are not interested in the future – dare I say not obsessed with the future – as "nostalgiacs" and "reversionists" who naively imagine a return to a mythical past.[17] Responding here to provocative questions posed by a Boston leftist publishing collective, he denounces as romantic antitechnologists the German "Greens," the "anti-nukes," and the other alleged reactionaries who align themselves with First Wave forces to try to prevent Toffler's future from emerging intact.

To be sure, in *Previews and Premises*, Toffler rejects the related notions that he is a technological determinist and that the future is predetermined. As he puts it, "I don't believe any single force drives the system. ... Different causal forces emerge as salient at different moments, and ... the attempt to find a single dominating causal force is a misguided search for a unique 'link that pulls the chain.'" Consequently, "the future is not entirely deterministic" as there is always the role of chance. "I certainly don't believe that the Third Wave is an inevitability." Not only can "the system ... go in any number of directions," but "all changes ... are made by people, including ordinary people, making decisions, choices."[18] This is considerably different from both *Future Shock* and *The Third Wave*, whether conceded by Toffler or not. The earlier supreme optimism has been shaken.

By 1990, with *Powershift: Knowledge, Wealth, and Violence at the Edge of the 21st Century*, Toffler has become even less optimistic. The ongoing, generally positive powershift throughout the world toward knowledge economies, decentralized governments, and participatory democracies is increasingly threatened by the possible rise of one or more racist, tribal (read nationalist), ecofascist, or fundamentalist states all too ready to suppress human rights, freedom of religion, and, not at least, private property. The broader context for this grave prospect is the continuing reapportionment of traditional authority throughout the world and the varying reactions to it. In tones reminiscent of *The Third Wave* he declares that "We live at a moment when the entire structure of power that held the world together is now disintegrating. ... we stand at the edge of the deepest powershift in human history," as if history really mattered to him. The proletariat is finished, replaced by the "cognitariat."[19]

Power, Toffler argues, comes in threes (like so much else) – violence, wealth, and knowledge – but has been steadily moving from the first two to the last. For example, the contemporary maldistribution of telecommunications facilities is allegedly more serious than our unequal food distributions. Violence and wealth alike, Toffler claims, are ever more dependent upon knowledge in order to be efficiently used or amassed. Traditional nations are steadily losing power to extranational forces ranging from multinational corporations, to ethnic groups, to organized religions, to criminal networks. The same technology that could achieve and sustain the glories of the Third Wave could instead be used to bring about a new Dark Age filled with violence as well as knowledge. Here Toffler is clearly correct if hardly original, as the 1989 and 1990 upheavals in China and Eastern Europe made abundantly clear. In any case, Toffler remains a virtual technological determinist, notwithstanding his 1983 disavowal.

Still, like all good jeremiads, *Powershift* holds out hope for silver linings if not outright redemption. Not only is the United States, along with Japan and reunited Germany, likely to dominate the coming international economy, but the United States is best positioned to remain the world's most powerful country, economy, and democracy – provided, of course, that the decision-makers read and accept Toffler's latest gospel. (Again, his 1983 disavowal aside, Toffler writes for elites, not ordinary folks.) The fact that the very global telecommunications and computer networks that Toffler routinely exalts could make books, above all big books, obsolete is an irony he ignores.

Despite his regular and well-publicized hobnobbing with the world's corporate and governmental elites, Toffler in *Powershift* as in *The Third Wave* wants to come across as a democrat, a friend of common citizens, and equally wants technology to promote democracy. He may genuinely believe that the sheer accumulation and generation of information through computers to unprecedented numbers of persons inevitably translates into a powershift from the centralized few decision-makers to the decentralized many. Ironically, he thereby reveals a principal failing of his book: its inability to prove that increased information necessarily means increased knowledge and so increased power. Toffler simply assumes this, when the compelling evidence is overwhelmingly negative. It is indicative of Toffler's "methodology" that he tries to distinguish among "data," "information," and "knowledge," only to give up and use them interchangeably, "even at the expense of rigor."[20] As before, he carefully avoids a serious critique of his *de facto* patrons, multinational corporations, as obstacles to true participatory democracy, focusing instead on easier targets like drug cartels, radical ecologists, and fundamentalists.

According to Toffler, *Powershift* is the final volume in his trilogy. *Future Shock* "looks at the *process* of change – how change affects people and organizations." *The Third Wave* "focuses on the *directions* of change – where today's changes are taking us." And *Powershift* "deals with the *control* of changes still to come – who will shape them and how."[21] This is a useful if oversimplified compartmentalization of his major works. One naturally wonders if he anticipated a trilogy when he began or if his works just conveniently became one. It will be interesting to see if, by the year 2000, Toffler nevertheless feels compelled to publish yet another tome and if such a work is as ambivalent and as anxious about the future as *Powershift* is, Toffler's surface optimism to the contrary.

John Naisbitt's *Megatrends: Ten New Directions Transforming Our Lives* appeared in 1982, two years after Toffler's *The Third Wave*. A self-styled "book of synthesis in an age of analysis,"[22] it professed to distill the insights about the near-future gleaned from twelve years' worth of 6,000 newspapers a month for a total of two million articles. Ironically, Naisbitt and his associates in the Naisbitt group use the very traditional methodology of content analysis. Notwithstanding the limitations of print in an age of broadcast media otherwise celebrated by Naisbitt, analyzing minor local newspapers' daily content supposedly reveals current trends, this on the dubious assumption that newspapers publish only those stories of most interest to their readers. What about their editors or their owners? And which comes first, public awareness or media coverage? In any case, the methodology transcends these questions insofar as more space in newspapers translates into more public concern, regardless of who saw the light first and who prompted its publication. In addition, the book's footnotes more often cite major newspapers and magazines than minor local ones, thereby relegating Naisbitt's data base to second place.

Like Toffler, Naisbitt uses many diverse events and developments that may seem random at first glance but that, not surprisingly, coalesce into identifiable megatrends. Naisbitt's acknowledged historical indebtedness is limited and dates back only to 1956–1957. By 1956, the United States had become a white-collar workforce of technical, managerial, and clerical personnel who outnumbered blue-collar workers, the traditional majority. And in 1957, with the successful Soviet launching of Sputnik, global telecommunications became a reality. The dual shift in the workforce and in communications ushered in the glorious information revolution.

Naisbitt's megatrends combine the obvious with the questionable. They include American movements (1) from an industrial to an information society; (2) from an internal to a global economy and marketplace, no longer producing and consuming most of what we need; (3) from northern to western and southern cities; (4) from short-term to long-term considerations and rewards; (5) from a representational to a participatory democracy; (6) from a hierarchical "top-down" to a "bottom-up" society; (7) from formal chain-of-command communications to informal communications networks in business, politics, and the home; and (8) from institutional to self-help in areas ranging from exercise and nutrition to job counseling. The other, no less important megatrends are (9) the matching of every successful high tech advance with "high touch," or a positive human response; and (10 the expansion of choices in everyday life from few if any to a multiplicity. That several of these megatrends are clearly applicable elsewhere in the world is, of course, a confirmation of their validity. It remains to be seen, however, if especially megatrends (4), (5), (6), (7), (9), and (10) apply as universally as Naisbitt contends, or just within the United States. High tech certainly has the capacity to reinforce or impose both hierarchy and traditional communications and to reduce choices. And high tech hardly guarantees "high touch" in any given situation.[23]

As more than one reverential reviewer observed at the time, Naisbitt was not a traditional futurist, offering "grand theories, imaginative projections, or reviews of past history."[24] (Toffler, for all his pretensions of uniqueness, *is* to that extent a traditionalist.) Rather, Naisbitt was describing, in his own words, "a new American society that is not yet fully evolved," but is "already changing our inner and outer lives"[25] – as if traditional futurists were not also invariably treating the present along with the future. But Naisbitt is again special, in the eyes of reverential reviewers, for simply telling "it like it is," "neither trying to predict the future nor convince the reader of a pet theory."[26] This focus on the present and near future, this dissociation among the past, the present, and the future, is a badge of honor for those concerned only with the short-term. Is that not contrary to one of Naisbitt's own megatrends?

Ironically, too, Naisbitt indeed has a pet theory beyond his true belief in free market capitalism as panacea: that once a megatrend develops, it will continue to develop, only more so. Here Naisbitt is anything but unique, as the historical study of forecasting's successes and failures makes painfully clear. Toffler, in fact, as noted, takes a similar stance. Still, the American tradition of positive thinking, of which *Megatrends*, along with Toffler's works, is only the high tech installment, prevails. Problems become challenges; challenges become opportunities. Millions of readers presumably share Naisbitt's conclusion: "My God, what a fantastic time to be alive!"[27]

But that was in 1982. By 1990, when Naisbitt and Patricia Aburdene published *Megatrends 2000*, things were looking even better. "We stand at the dawn of a new era," they claim at the outset. "Before us is the most important decade in the history of civilization, a period of stunning technological innovation, unprecedented economic opportunity, surprising political reform, and great cultural rebirth. It will be a decade like none that has come before because it will culminate in the millennium, the year 2000."[28] Once again the past supposedly has no hold on the present, much less the future.

Naturally there are ten new megatrends to ponder and then act upon: (1) a global boom free from past limits on growth and without any future limits; (2) the emergence of "free market socialism" (an oxymoron) in Eastern Europe; (3) the privatization of the welfare state as in Britain under Margaret Thatcher; (4) the rise of the "Pacific Rim" extending from California to every other country, including those in South America, fronting on the Pacific; (5) global lifestyles and cultural nationalism, including the attempted preservation of national or regional cultures amid global homogenization; (6) a renaissance in the arts so popular as to replace sports as some societies' dominant leisure activity; (7) unprecedented leadership of women in business and the professions; (8) worldwide religious revivalism from fundamentalism to New Age; (9) an age of biology and biotechnology (the only megatrend that troubles them ethically and otherwise); and (10) the triumph of the individual over the collective through high tech computers, cellular phones, and fax machines. *Megatrends 2000* is more international in scope than its predecessor. As *Megatrends 2000* happily concludes, "On the threshold of the millennium, long the symbol of humanity's golden age, we possess the tools and the capacity to build utopia here and now."[29]

Paradoxically, Naisbitt and Aburdene claim to be truly forecasting, if only for a decade. But their forecasts here are more obvious, less imaginative and insightful, and even less based on compelling hard data than those in *Megatrends*, which avowedly steered away from forecasting. Virtually all of their major points have already been detailed in such periodicals as *Time*, *Newsweek*, *Business Week*, and the *Economist*, while megatrends (6) – renaissance of the arts – and (10) – the triumph of the individual – surely remain to be confirmed. Global environmental crises, hunger, crime, drugs, AIDS, and other genuine crises are either ignored or minimized. While *Megatrends* at least appeared when most of the developments it described were not yet part of the conventional wisdom, *Megatrends 2000* is behind the times, so to speak, in its revelations. Even so, Naisbitt and Aburdene defend both books by arguing in the sequel that *Megatrends*' trends are on schedule but that the latest shift is the quickened pace of information growth, thereby requiring *Megatrends 2000*.

Significantly, Naisbitt and Aburdene's conclusion also includes the following statement: "By identifying the forces pushing the future, rather than those that have contained the past, you possess the power to engage with your reality."[30] Once again history is irrelevant save as the supposed contrast with the golden age ahead. It is as if high tech arrived and flourished in an historical vacuum of no more than a few decades and as if everything before it can simply be forgotten. Such historical amnesia, combined with Naisbitt and Aburdene's dismissal of caution on the road to utopia and their contempt for "naysayers," reveal their unacknowledged anxiety that history in some fashion may indeed repeat itself; that technological progress may once more be a mixed blessing; and that unanticipated trends or events may wreck or detour their smooth scenarios. It is precisely when American business and professional people, among others, are so unsure of the country's or the world's future, so desperate to be handed sugar-coated soma pills in book form, that *Megatrends 2000*, like *Megatrends*, can be so successfully packaged, purchased, and consumed.

What, however, is most troubling about Toffler, Naisbitt, and like-minded high tech prophets is the virtual absence of any moral critique of the present, any deep-seated concern driving them to their speculations about tomorrow. Even Buckminster Fuller or Gerard O'Neill, with their naive beliefs in the prospect of achieving versions of technological utopia either on spaceship earth or in space colonies, had a fundamentally nonpecuniary interest in the future. Both wrote and spoke out of conviction, however misplaced, that they had panaceas worth putting into effect. In this respect they continued the tradition of utopian – and, equally important – antiutopian writing and organization dating back at least to Thomas More if not to Plato. With Toffler and Naisbitt, by contrast, one senses little more than crass opportunism, including a willingness to revise radically one's predictions to meet the marketplace, and absolutely no sense of being part of any historical movement. It is no accident that Toffler and Naisbitt alike praise "wealth creation" (Toffler's term) as mankind's greatest achievement. It is something they do know about.

II

By contrast, high tech print and television advertising boldly uses historical figures and structures to promote its products but in so doing both misappropriates and trivializes the past and bespeaks an anxiety about the present and the future akin to that of Naisbitt and Toffler. The most provocative examples, though hardly the only ones, are Xerox's use of Leonardo da Vinci; Apple Computer's use of Isaac Newton; Bell South's use of Ralph Bunche, Winston Churchill, and Albert Schweitzer; IBM's use of Charlie Chaplin's Little Tramp; and the modeling of Dallas' Infomart, housing information processing systems and services, after London's 1851 Crystal Palace, site of the first world's fair.

As a student of American culture, I am hardly surprised by the use of major historical figures or structures to increase sales and profits. Every February, for instance, I anticipate ever cruder commercials proudly showing George

Washington and Abraham Lincoln – our apostles of truth – as happy salesmen for items ranging from cars to televisions to dishwashers. I do wonder whether their association convinces even a single potential consumer to make a purchase. Likewise, I am no longer shocked by the association of, say, the Statue of Liberty or the Liberty Bell with commercial wares having absolutely nothing to do with either's history and significance. Yet these associations, however crass, are not a recent development and at least stir a bit of genuine nostalgia for "the good old days," variously defined, when America's leaders and symbols were more uplifting and respected than in recent years.

As a student of technology, however, I am disturbed by the false nostalgia created when high tech companies and institutions – or their advertising agencies – adopt wholly inappropriate persons and buildings to push their products. Take the case of Infomart, which opened in 1985 at a cost of $100 million.[31] Just as London's Crystal Palace showcased the products of the British industrial revolution, Infomart is supposedly just doing the same for those of high tech. Its designers, we are told in an early promotion brochure, "have meticulously adapted [Joseph] Paxton's original plans for the Crystal Palace. ... Great care has been taken to maintain the historical integrity of the original structure while designing a state-of-the-art facility."[32] A 1992 promotion publication goes even further. Infomart has somehow "been recognized by Great Britain's Parliament as the official successor to the Crystal Palace. It reflects not only the Crystal Palace's architecture, but its spirit and forward-thinking purpose."[33] Moreover, the same Italian company that designed the Crystal Palace's fountain, Barovier-Toso, designed Infomart's crystal fountain; and one company that exhibited at the Crystal Palace, Siemens and Halske Telegraph, then manufacturing railway telegraph and signal devices, has a showroom at Infomart as Siemens Information Systems, now manufacturing automation devices. But not only is the 1.6 million square-foot, eight-story Infomart almost one and a half times the size of the Crystal Palace, both its exterior and interior are hardly identical with the latter, as befits an ultra-modern convention and exhibition facility. Inside, it more closely resembles the kind of multi-story atrium found in architect John Portman's hotels. A more successful contemporary structure is New York City's Javits Convention Center, a public facility that also has ties to the Crystal Palace, but less explicit and less pretentious ones.[34]

More important, Infomart's purpose is not to generate world peace, or serious cultural exchange, or even wholesome amusement, as was true of the Crystal Palace, but only commerce. The presence of a high tech electronic library and resource center and of endless seminars, demonstrations, and showcases, and the absence of sales people, may indeed provide a more relaxed, more "user friendly" atmosphere conducive to comparison shopping; but selling the products and services displayed, not satisfying intellectual curiosity, remains Infomart's sole objective. Boldly describing Infomart as "the education center for the Information Age"[35] hardly alters that fact. In a world of instantaneous communications and "electronic cottages," as elaborated below, it is no longer necessary, as was the case with the Crystal Palace and other pre-1930 world's fairs, for business and professional people to come to one central location to examine the hardware and software; they can remain in their home offices and have the information come to them through their computer and television screens and fax machines. Ironically, the same high tech firms leasing space in Infomart are those who make just such claims in other contexts, such as print and media advertisements. In fact, Infomart's own quarterly magazine has a recent article praising video-conferencing for doing precisely this and so eliminating unnecessary business and professional travel to distant sites![36] Ironically, too, the Dallas Infomart was intended as just the first in a series of Infomarts yet to be constructed in New York City, Atlanta, Chicago, Los Angeles, Frankfurt, and elsewhere; one in Paris is now open. So much for literal adherence to the one-place-at-one-time centralization of exhibits in traditional world's fairs. In addition, rival computer marts are in the works.

If Infomart is a cultural symbol at all, it surely represents high tech's lack of self-confidence, its ambivalence about the present and the future, and its consequent need to establish close ties to the past. Infomart's intriguing facade thus remains merely a facade. Only insofar as it houses an equally new Information Processing Hall of Fame, including John Cullinane, Grace Hopper, H. Ross Perot, and An Wang, has Infomart any historical or cultural bent. And the idea of a hall of fame in technology perpetuates the simplistic, romantic notion of an "heroic theory of invention" generally rejected by serious students of technology's past.

Despite all this, Infomart at least has its commercial connection with the Crystal Palace, which indeed had business high on its agenda. By contrast, Leonardo da Vinci (1452–1519) has no ties whatsoever with high tech save as a pioneer designer and inventor, among other talents.[37] Somehow we are supposed to see this individual genius as a loyal corporate employee, happily embracing photocopiers, fax machines, laser printers, and entire workstations. Presumably such high tech equipment would have enabled Leonardo to design, invent, sketch, and even paint better. Yet the Xerox ads sometimes show Leonardo as a befuddled time-traveler unable to cope with the modern office. The tag line that "Xerox brings out the genius in you" may, then, be an implicit criticism of eccentric souls who could not last a day in the modern high tech corporation – not on "Team Xerox" anyway – as much as an obvious paean to the hardware and software themselves. The tag line is certainly an implicit relegation of Leonardo's intellect to second place behind the high tech equipment he now hawks. Once again, though, high tech feels it necessary to look backward for reassurance as it otherwise looks only ahead.

By coincidence, in the early 1980s IBM (and Bechtel) sponsored a traveling exhibit, "Leonardo's Return to Vinci," that toured several of America's university art museums. Its ulterior purpose was probably lost on most viewers dazzled by the actual art and models of inventions (tank, parachute, helicopter, car, paddlewheel ship, etc.) of this archetypal Renaissance man: to demonstrate that high culture and high tech mesh as readily in our time as in Leonardo's.[38] Or, as a newspaper article stated at the time, if Leonardo were alive today he would contentedly be working (if not at Xerox then) as an artist in the electronic medium, with the TV screen as his canvas and with knobs and computer keyboards as his brushes and paints.

Ironically and unintentionally, the exhibit demonstrated the very opposite, and in ways that were as revealing as the displays themselves. The truth about Leonardo is that his age – unlike ours – tolerated and actually encouraged the diverse activities that made him a Renaissance man beyond chronology alone. Moreover, notwithstanding his individual talents, he was hardly unique in his time. Attributing Leonardo's manifold achievement in painting, sculpture, design, architecture, and engineering solely to his singular genius is to miss the point: he was multifaceted in part because there did not then exist both the tremendous gaps between technology and culture and the occupational specialization that we today accept as almost inevitable. Technology in its present meaning also did not then exist, and the architect and engineer (if not the artist) were invariably the same person.

Leonardo would not, of course, have fit the conservative corporate molds of IBM or Bechtel any better than that of Xerox. Nor would any of these three corporate cultures have granted him the relative independence he enjoyed in his own day. Still, among these corporate sponsors there likely lurks a hope that similar inventive geniuses will reappear to rescue the United States from its present technological and economic malaise and that corporate America will somehow find a way to accommodate them. The traditional American admiration for alleged lone-wolf inventors like Eli Whitney, Henry Ford, and Thomas Edison persists. Like them, Leonardo actually invented considerably less than he is credited with; and, like them too, what he did "invent" – at least in his sketches – was frequently the result of collaborative efforts.

Persons of Leonardo's nevertheless impressive intellectual stature do, of course, occasionally appear, but they no longer resemble our Renaissance man. That is the underlying, the unintentional message of this exhibit and of the Xerox advertisements: that the past is not recoverable by the present. Or, to quote, of all persons, the operations manager of Xerox's facility of Dallas' Infomart, praising Infomart's integration of goods and services, "I imagine it's very annoying to see technology out of context."[39] Indeed! High tech will have to look elsewhere for its cultural heroes.

This Renaissance pseudo-heritage came to haunt high tech when, on March 6, 1992, a new computer virus named Michelangelo struck at least five large and a thousand small businesses in the United States and an untold number of individuals. Given high tech's self-proclaimed universality, the virus also struck around the world, infecting an estimated five million personal computers. The occasion was the Renaissance painter and sculptor's 517th birthday, and among the American high tech firms affected was, most ironically, the DaVinci Systems Corporation, a maker of electronic-mail software. It had recently mailed out some nine hundred infected demonstration disks. Other software and hardware manufacturers were also victimized by the virus, a program planted secretly by one or more malicious programmers. Hidden on infected floppy computer disks, the virus spreads to (in this case IBM-compatible) computers when the infected disk is used to start (or boot) them up. Once this happens, the virus program writes itself into the computers' memory and infects the computers' internal hard drive, or memory device. And if the infected floppy disks are used in other computers, the virus will spread to them as well. The result is the loss of invaluable data often not backed up on either additional disks or hard copies. Lying in wait for its potential victims, the Michelangelo virus was triggered, as expected, on March 6, 1992, or whenever infected computers' calendars registered that date. True, well-publicized warnings and the availability of software that detects and removes viruses greatly reduced the damage. And antivirus software manufacturers and their advertisers benefited tremendously. Yet every March 6 for the foreseeable future Michelangelo the Renaissance artist and humanist will have to compete with Michelangelo the destroyer of knowledge.[40]

On a more positive note, perhaps, we can anticipate any number of charming advertisements when Apple Computer begins selling its recently unveiled Newton, a pocket-size, pen-based computer named after scientist Isaac Newton. The first in a projected line of "personal digital assistants," the Newton combines an electronic calendar, a card index, a note taker, and other personal organizing functions. Whether this device will prove commercially popular, especially with nonusers of conventional computers, remains to be seen, as does – more pertinent here – the role of that preeminent scientist in promoting it. Newton is widely associated with apples in his experiments on gravity, particularly gravity's extending to the moon, but its association with his discoveries about gravity may have been his own clever invention, intended to backdate those discoveries to establish priorities over rival claims. In any case, there is no logical connection between those seminal discoveries about gravity and Apple Computer's Newton. The latter's incorporation of "smart agents" or "intelligent assistance," software terms referring to the organization of information without computers being told to do so, is the only conceivable tie – that is, both Newtons being terribly bright. (Whether, like Leonardo in the Xerox advertisements, the original Newton's intelligence will be deemed inferior to that of his high tech namesake will be interesting to observe.) In Isaac Newton's day, science and technology were separate realms with different agendas, unlike, as with Leonardo, architecture and engineering.[41]

By contrast, there is no tie whatsoever between Ralph Bunche, Winston Churchill, and Albert Schweitzer and the components of Bell South, one of the "Baby Bells" created by the court-mandated breakup in 1984 of American Telephone and Telegraph. The fact that each of these persons had multiple talents – athlete, college professor, author, statesman, and negotiator (Bunche); war correspondent, author, painter, and world leader (Churchill); theologian, physician, organist, author, and humanitarian (Schweitzer) – is somehow supposed to place Bell South itself in a new light: "everything you expect from a leader" in telecommunications, information services, mobile communications, and even advertising. This crass association of the allegedly "most admired telecommunications company in the United States" with three individuals hardly revered for anything technological and having precious little in common with one another is historical stretching of the worst kind. It is compounded by the ridiculous notions that the "remarkable talents" found in each man can be molded to fit conservative corporate cultures, much like Leonardo's creativity, and that strong-willed individuals like these can be inspired to work together to solve problems. What compels Bell South to look back to such irrelevant historical figures to sell its products?[42]

Perhaps the most offensive high tech advertisements, however, are IBM's caricatures of Charlie Chaplin's Little Tramp.[43] These helped launch IBM's personal computer in 1981 and ran until 1987. They were revived in 1991 to celebrate IBM's tenth anniversary in the personal computer business. Cleverly playing on the pervasive fears of modern technology itself creating chaos – not least on the assembly line, as epitomized by Chaplin's classic 1936 film *Modern Times* – the IBM advertisements instead show a benign view of personal computers, as devices bringing order, profits, and happiness into the Little Tramp's previously disorderly, inefficient, and unhappy life. Whether the setting is a house, an office, a bakery, or a hat shop, the outcome is always the same and is always happy, supposedly in accordance with the film's ending and view of life. Indeed, several IBM advertisements call the personal computer "a tool for modern times," and the film, not just the Little Tramp, was clearly on the minds of the ad campaign's creators. At the same time the advertisements humanize the world's preeminent computer company, the epitome of the impersonal large corporation. In so doing they completely misrepresent Chaplin's critical views of modern technology and its social impact and of the large corporations responsible for so much of both. Contrary to the IBM advertisements, *Modern Times* finally seeks escape from modern technology and industrial society and, if anything, is wholly inadequate in not suggesting any (other) solutions to the dehumanization of work and leisure it depicts. If only the Little Tramp had had a PC at hand! Ironically, the film's other serious message beneath its comedy is that human beings cannot master modern machines. Like the recent television commercials for luxury cars, financial securities, jogging accessories, and other "upscale" goods that use popular songs – and sometimes singers – from the 1960s with originally antimaterialistic messages, the IBM advertisements create a particularly distorted view of the past, an utterly false nostalgia.

Ironically, high tech *has* a history or is creating one with the several "generations" of computers that have come into existence since World War II – the term "generations" itself a telling comment on the rapid pace of change within high tech. Appropriately, the computer now has its first major museum, in Boston, housed in what a century ago was a warehouse. Here there is no contradiction between the building and its contents, for the structure, despite the parallels between the earlier industrial revolution and the present information revolution, lacks the architectural and cultural pretensions of Infomart. Meanwhile the museum, being a nonprofit enterprise, lacks commercial aspirations and has recently and significantly modified its initial basic message of computers' evolution as constituting unadulterated technological and social progress.[44] Infomart's Information Processing Hall of Fame would be inappropriate in the revamped Computer Museum. Would that Infomart, Xerox, IBM, and other high tech institutions ever felt secure enough to present a more balanced picture of modern technology's mixed blessings. Their doing so would be a reflection of high tech's genuine self-confidence.

It is, of course, entirely possible that these and other high tech companies and their advertising arms care not a wit about historical accuracy and happily misuse historical figures and structures just because they sell their products

– or seem to sell their products, given the notorious difficulty of connecting advertisements with sales. Or, for all I know, these companies and advertisers may be blithely unaware of their perversion of history and may believe that they are somehow promoting history in a manner akin to corporate sponsorship of other historical and cultural enterprises. Still worse, these companies and advertisers may, as with the IBM Little Tramp ad campaign, apparently believe that their historical figures would have embraced their high tech products if only they had seen them in action; that Chaplin himself would have been genuinely liberated by a personal computer and would have become a contented entrepreneur. The linkages between high tech and high culture are sufficiently unexplored to allow for all these possibilities.

There are, in fact, other high tech companies who utilize historical figures and buildings in more historically responsible advertisements (if advertisements for commercial products can ever be historically responsible). These include Sun Microsystem's adoption of Mozart at the piano to boast that "only one other ten-year-old has ever performed so well on a keyboard" on the computer company's tenth birthday; Storer Communications' comparison of a nineteenth-century telegraph operator who "could tap out about twenty-five words in a minute" with computers that "can deliver a 300-page report in seconds"; Storer's similar comparison of nineteenth-century Pony Express riders through whom "communications traveled about fifteen miles per hour" with contemporary systems through which "communications can travel at the speed of light"; and Lockheed's use of William the Conqueror's mounted cavalry on stirrups against the English Anglo-Saxons in 1066 to proclaim its own "air superiority" in the then ongoing Cold War against the Warsaw Pact's aircraft.[45] In addition, the El Al Airline bimonthly magazine I read while going to the conference where this article was originally delivered contained an advertisement for an Israeli tool manufacturer, Iscar, comparing the art works of Michelangelo and Leonardo, among others, with its own – its hardmetal cutting tools "which grace no famous museum but are utilized regularly in industrial sites around the world."[46] An audacious comparison, perhaps, but a valid one insofar as art and (what we call) technology historically were inextricably connected and often equated. In 1829, Harvard professor Jacob Bigelow's *Elements of Technology*, which, along with his lectures, largely introduced the term in the United States, defined technology as the "application of the sciences to the useful arts."[47] "State of the art technology," a term invoked so frequently today by high tech promoters, thus had a deeper meaning than imagined by nearly all who employ it.[48]

Yet the more widely promoted misuse and, equally important, trivialization of avowedly humanistic symbols like Leonardo and the Little Tramp by Xerox, IBM, and others cannot be so easily excused as oversight or indifference or good intentions gone wrong. One need not be conspiratorial to recognize other forces at work and to suggest high tech's need to connect to the past. If, then, high tech feels compelled to look backward for reassurance or legitimacy – or simply greater profits – let it be more honest in its historical appropriations. Otherwise, its false nostalgia may someday prove self-defeating.

III

High tech increasingly finances world's fairs – another form of advertising – but in so doing fails to achieve the intellectual significance of prior fairs and raises questions about the continuing relevance of world's fairs. This heretical notion first occurred to me at, of all places, a 1980 symposium commemorating the 1939–1940 New York World's Fair held on the very site of that fair, Flushing Meadows (also the site of the 1964 New York World's Fair). Although other international expositions from 1851 onward displayed no less impressive exhibitions than New York's did, the 1939–1940 fair alone announced the prospect of creating a veritable utopia in the very near future: in 1960, to be exact. Its achievement quite literally awaited the magic touch of the fair's several prominent industrial designers: Norman Bel Geddes, Henry Dreyfuss, Raymond Loewy, and Walter Dorwin Teague.[49] Or so they – and thousand of fair visitors – naively believed.

As with so many other technological utopias – and not only world's fairs but also model communities and visionary writings – the problem has been twofold: the inability to predict the "real" future technologically and nontechnologically, and the inability to translate actual technological advances into equivalent social advances. The same problem, of course, plagues high tech's prophets. By 1960 American society resembled the "world of tomorrow" only in bare outline – in its sleek skyscrapers and superhighways. Much remained to be filled in, and obviously still does.

Among planners of recent and coming world's fairs both at home and abroad there seems painfully modest concern for the future of fairs as significant social and cultural artifacts. Like Toffler and Naisbitt, these high tech visionaries are ahistorical, so ahistorical as to fail to ask themselves how past world's fairs have affected those who created, visited, and read or heard about them. Yet this is precisely where world's fairs are – or at least have been – most significant. They have revealed much more about the times and places that produced them than about future times and places. As scholars in several disciplines are gradually recognizing, fairs embody the real or ideal self-image held by their respective organizers and exhibitors and so provide a wonderful opportunity to probe deeply into any number of societies and cultures on display. This is, if anything, more important today than ever before just because the alleged shrinkage in time between the problematic present and the future perfect – as exemplified in Buckminster Fuller's persistent demand that we either create utopia immediately or face oblivion – likewise lessens the role and excitement of world's fairs as previews of the future. If our high tech future is virtually at hand, what can fairs actually teach us about the world of, literally, tomorrow?

The historic importance of world's fairs thus does not guarantee their continuing importance. Apart from the always sensitive question of finances, given their common failure to turn a profit, there is a no less weighty question of the continuing utility of world's fairs as conveyers both of ideals and of technical information. Just as postage stamps, peace ships, and peace congresses are no longer viewed as efficient means of achieving international harmony, world's fairs ought not to be so viewed. Simply bringing together masses of people into one geographical space is hardly a serious route to that admirable goal. Other means to world peace more suitable to the late twentieth century should be sought.

Technology comes readily to mind, whether as a military deterrent or in more positive forms. Yet it is the advance of technology since 1940 that, more than anything else, has rendered fairs obsolete. The revolution in electronics and information processing, not even envisioned in 1939–1940, has made possible instantaneous visual and other communication throughout most of the globe. Why travel to distant points if satellites, computers, word processors, and other high tech advances can both deliver the needed data and enable one to see the latest technological developments on a screen or a fax rather than in person? Those who still wish, and can afford, to view technological developments in the flesh can of course do so, perhaps at the local, regional, national, and international trade fairs that persist today; or at Dallas' Infomart and similar structures elsewhere. But save for Infomart, these enterprises have no serious social and cultural pretensions, even those that call themselves a "World's Fair for Consumer Electronics" or a "World's Fair of Imaging."[50] Hence the other principal purpose of international expositions – bringing technological advances to the attention of the largest number of people in the most effective way – has likewise been severely undermined.

True, world's fairs have traditionally promoted entertainment and good times as well as peace and technology. And many medieval precursors of the Crystal Palace mixed trade shows and carnivals. It is hardly accidental that amusement parks are more commonly associated with world's fairs for many persons than anything else. Yet here, too, international expositions are obsolete. They have been replaced by permanent "theme" parks exemplified by Disneyland and DisneyWorld, which are the entertainment counterparks of Infomart. Theme parks are now the rage throughout the world. They increasingly encompass every conceivable popular activity and fantasy, from computerized high tech roller coasters and other thrilling rides, to movie studio recreations, to foreign travel, to themes like the sea, country music, sports, cars, toys, and celebrities. Yet the opening in 1983 of Tokyo Disneyland and in 1992 of Euro Disney near Paris should be construed as the triumph not of true internationalism but of American cultural imperialism. Calling Euro Disney "a cultural Chernobyl," as at least one widely quoted French intellectual did, is, however, going a bit too far.[51] These new Disneylands are more akin to the American fast-food eateries that now dot so much of the "global village." They carefully blend popular features from their California and Florida counterparts with appealing aspects of Japanese and French culture respectively but remain fundamentally American in character and content. That they are so popular with the Japanese and the French does not contradict this. Their popularity ironically lessens the motivation for non-Americans to visit Disneyland and DisneyWorld and to this extent undermines the internationalism celebrated by traditional world's fairs and by some theme parks.

In the case of Disney's EPCOT center, visitors are provided with an even more romanticized vision of both the past and the future than temporary fairs once offered and as strong a dose of technological determinism as in Toffler's and Naisbitt's writings. Officially named Experimental Prototype Community of Tomorrow, EPCOT was envisioned by Disney as a dome-covered actual city of 20,000 full-time residents. Before he died in 1966, Disney left few specifics as to how his technological utopia should look and operate. Eventually his successors decided to do without people – perhaps the ultimate high tech dream – and to create, adjacent to DisneyWorld, exhibits divided into World Showcase and Future World. World Showcase offers replicas of buildings from nine nations (plus the United States) to allow visitors the opportunity to experience foreign lands without leaving American soil. It is as artificial and as superficial as any traditional world's fair, or perhaps worse, for in EPCOT other countries do not display their latest inventions, only corporations do.[52]

Future World exalts the ability of its giant corporate sponsors to solve problems confronting mankind through more and more technology. The notion of corporate responsibility for any problems is naturally left out. Moreover, problems that ordinary citizens might deem serious, much less fundamental, are conveniently ignored: war and peace, overpopulation, poverty, unemployment, crime, drugs, etc. Here as elsewhere in the Disney empire, painful reminders of everyday life are, along with each day's litter, religiously swept away. Ironically, visitors to EPCOT and to DisneyWorld have increased so much since the former's 1982 opening that they are often confronted with real world problems upon arriving and departing if not during their stay, from poorly planned surrounding areas to inadequate public transportation.[53] Even more ironically, few of the Future World exhibits treat the future; most celebrate existing and past technologies. Robots act out everyday scenes from prehistoric times on. The General Motors pavilion includes the model for Leonardo's *Mona Lisa* tapping her feet and scowling while our Renaissance hero works on his flying machine. So here Leonardo sells cars, in effect. One could hardly ask for clearer confirmation of high tech's inability to escape the past in the name of promoting the future.

Significantly, EPCOT does not merely sanitize the past, in the manner of other temporary world's fairs and permanent theme parks, but outright "improves" it. History as presented here is devoid of what Disney employees call "downers," such as famine, plague, or genocide. Like DisneyWorld's Main Street, it is history as it should have been or should be. The American pavilion, sponsored by American Express and Coca-Cola, uses robots dressed as Benjamin Franklin and Mark Twain to romp through American history in twenty-nine minutes. Wars, depressions, and suffering are not completely left out, but the overall thrust is naturally upbeat. Coverage of the 1960s and 1970s, for instance, totally ignores race riots, Watergate, the feminist movement, and Vietnam, among other unpleasantries. Lest even this tax visitors, there is, by EPCOT's own admission, World Showcase to relieve information overload.

As historian Mike Wallace has put it so well, EPCOT overall "promotes a sense of history as a pleasantly nostalgic memory, now so completely transcended by the modern, corporate order as to be irrelevant to contemporary life." Hence the tone of gentle mockery and wry amusement in the Future World reconstructions of past futile efforts of lone inventors and small businessmen to satisfy consumers because naturally only large corporations can successfully manage this nowadays. That in turn "dulls historical sensibility and invites acquiescence to what is" or what is soon coming.[54] This distortion and detachment of history applies, of course, to high tech prophecies and advertisements.

EPCOT is praised by Disney executives and (other) public relations professionals as a serious educational enterprise, testing ideas and inventions, not merely another Disneyland. The lessons learned, however, do not, contrary to

Walt Disney's dream for EPCOT, begin to apply to solving urban problems. Like so much else of high tech, EPCOT is an escape from reality, not an illumination of it. As a senior Disney executive confessed in 1982, "We're interested in seeing technology work to accomplish a story point," not quite the same as a practical solution to problems. "We wanted to make a point about America, that dreaming and doing things is an ongoing thing." This is very different from showing how things actually work, much less how they came about. EPCOT, he continued, is a necessary "voice of optimism" amid mounting uncertainties. "Industry has lost credibility with the public, the government has lost credibility, but people still have faith in Mickey Mouse and Donald Duck."[55]

Going a step further from urban problems, Hollywood's Universal Studios, already operating a Florida theme park competing with DisneyWorld, is completing a four-block entertainment and retail area in Los Angeles, called City Walk, that recreates a Los Angeles that never existed, then terms that pseudo-past the wave of the future. Combining shops, restaurants, and clubs amid a mixture of such architecture styles as Art Deco, Cyberpunk, and Southwestern, City Walk provides a "safe, thematically correct" locale for pleasurable activities free from reminders of Los Angeles' more mixed and less appealing actual development. City Walk's features include a replica of a giant surfboard as one building's roof, a section of pavement embedded with fake litter, and street performers. The absence of a sustained sense of history often attributed to Los Angeles – along with the scant number of actual places like City Walk to gather and to walk – allegedly justify this historical "reconstruction." According to a project architect, City Walk is the "architectural equivalent of artificial insemination." Its ultimate objective is to "jump-start" the local culture "by infusing it synthetically and letting it evolve from here."[56] This might become the goal of other theme parks as well.

Missing from these theme parks and recent world's fairs as from Toffler's and Naisbitt's writings is any genuine moral critique of the present, any serious effort to alter society in the manner of the 1939–1940 New York and other major pre-World War II fairs. As socially and culturally conservative as those earlier fairs admittedly were in their common embrace of corporate capitalism, not to mention their implicit sexism and racism, they did offer a view of the future intended to inspire change, not simply to entertain. And it is increasingly difficult, perhaps impossible, for world's fair organizers to match the expertise and imagination of theme parks' professional entertainment specialists and designers; or to avoid hiring them in order to compete for a season or two with the Disneylands and EPCOTS.

These contemporary professionals, moreover, generally lack the cosmopolitans of a Bel Geddes or a Loewy, for whom fairs were but one of many diverse projects rather than preoccupations. They did not spend their time attending meetings of the International Association of Fairs and Expositions, now a quarter-century old.[57] It is revealing of such groups' ahistorical perspective that their very invocations of history, like those of Toffler and Naisbitt, are at best embarrassingly shallow and at worst plain wrong. At the 1980 symposium commemorating the 1939–1940 New York fair, for example, the only participant who appeared fully confident that the future really could still be fashioned and improved in the manner of the "world of tomorrow" was such a professional world's fair organizer. His talk on "From Out of History Comes Energy Expo '82 – the Knoxville World's Fair" demonstrated absolutely no grasp of the historical discontinuities that had already made the Knoxville fair obsolete. For him, as for Toffler and Naisbitt, the past was a convenient if barely acknowledged launching pad for the future, and whatever trend or wave currently existed – in this case, ever more world's fairs celebrating one technological achievement after another – presumably would continue indefinitely. Similarly, the head of marketing and communications for Vancouver's Expo '86, a much more successful world's fair than Knoxville in strictly quantitative and financial terms, boasted in a subsequent trade publication interview of "How PR Packed 'Em In At Expo '86" by promoting it as "a world exposition in the grand tradition." His historical knowledge, alas, was limited, as per his muddled assertion of "trying to reference to the people and the media that we were going back to what it was like in London, and what it was like in 1861, when the very first world exposition took place." That he presumably meant 1851 was only the more obvious of his errors, the more serious of which was the assumption that what life was like at the 1851 Crystal Palace could readily be duplicated in 1986 (and then, of course, surpassed).[58]

Finally, the principal sponsorship of world's fairs has, like so much else, gradually shifted from the public to the private realm. As fewer and fewer cities and states and even nations have been able to afford to spend enormous sums on world's fairs, and as Chicago, Oklahoma City, St. Louis, Paris, and Venice have consequently cancelled long-anticipated fairs, giant high tech multinational corporations have taken their place.[59] To take the two most recent American fairs, the federal government contributed $200 million to Knoxville's 1982 extravaganza, but gave only 10 million to New Orleans' 1984 effort. The American pavilion at Seville's 1992 world's fair was nearly left unfinished because of allegedly inadequate government funding. Seville, in fact, marks the first time that the federal government, under intense Congressional pressure, has sought private sector funding for its world's fair pavilion.[60] This is considerably different from earlier times, when major corporations either displayed their products in government-sponsored exhibition halls or, when constructing their own halls, did not dwarf those built by the many cities, states, and nations also in attendance. It is no accident that Dallas' Infomart bills itself as the "largest privately owned exhibit hall in the world" and deems that a mark of prestige.[61]

Moreover, and more so than in the past, these contemporary corporations treat world's fairs, like theme parks, as little more than huge advertisements for their products; and if world's fairs are to survive, they will assuredly have to contribute substantially to the "bottom line." The legacy of the Knoxville and New Orleans fairs, not only in losing tens of millions of dollars but also in leaving unoccupied and unusable buildings, would hardly be tolerated by profit-minded corporations. The example of the futuristic United States exhibition hall in Knoxville standing vacant and deteriorating would tarnish the image of any self-respecting corporation with a similar white elephant, while the pre-closing declaration of bankruptcy by the New Orleans fair organization would obviously be the height of embarrassment for any corporation in a comparable situation.[62] For these reasons it is increasingly common for corporations to promote their products as official fair suppliers (clothing, for example) or representatives (airlines, for instance) rather than, as was done earlier, as outright exhibitors. This in turn gives new meaning to the phrase "selling the future," which is associated with world's fairs, theme parks, and related enterprises.[63] Not surprisingly, the author of a disappointing study comparing the "participant outcome perspective" of visitors to the Knoxville and New Orleans fairs concludes with a plea to increase "advertising and marketing events to tell the [innocent?] individuals what to expect from the experience" and to provide assistance "to interpret and understand their experience on-site."[64] Left to their own devices, visitors apparently might get the wrong message about the future and, more important, about its corporate sponsors.

Pleading, in November 1991, for the $6 million then needed to complete the underfunded United States pavilion for the Seville world's fair opening the following April, one Henry Raymont wrote in the *Boston Globe* that most Americans failed to grasp the enormous symbolic significance of an appropriate American presence there. The problem, he contended, was not "partisan politics or genuine budgetary concerns" so much as "a lack of historical imagination."[65] As the chairman of the Culture Committee of the Christopher Columbus Quincentenary Jubilee Commission, Raymont naturally had a vested interest in having the pavilion – already criticized as consisting of little more than two recycled geodesic domes used in European trade fairs over the last twenty years – completed on time. Yet the irony of his historical reference cannot go overlooked, not least because, for political reasons above all, what began as a celebration of the 500th anniversary of Columbus' first voyage to the New World, ended, by the host country's own doing, as a celebration instead of Spanish democracy, culture, and technology. Spain's king did not even mention Columbus' now controversial name when he officially opened what has been billed as the biggest world's fair ever, in terms of the number of countries represented and of exhibits constructed and of costs incurred.

If, for once, the United States was not the biggest exhibitor, that could be the healthy start of rethinking the continuing significance of world's fairs in the age of high tech. Contrary to Raymont and other defenders, the same technological progress that formerly inspired so many fair designers and patrons has rendered the object of their affections irrelevant to the future. At best, fairs are historical artifacts, beloved by nostalgia buffs. Nowhere among recent fairs has this point been demonstrated better than at Vancouver's Expo '86, whose very theme was "World in Motion, World in Touch." The communications and transportation technology most prominently displayed repeatedly if unintentionally confirmed the technological obsolescence of world's fairs. The fact that the Vancouver fair's overall message was that technology cannot solve all world problems only compounds the irony, as does the fact that visitors dazzled by the high tech displays rarely got this message. No other recent fair has shown such courage in suggesting modern technology's mixed blessings, and even Vancouver lacked the "historical imagination" to illuminate historical developments leading to the end of technological optimism.[66] Far more common is the message at a major pavilion at the 1985 Tsukuba, Japan, world's fair, whose overall theme was unadulterated technological progress: "What mankind can dream, technology can achieve."[67] As with high tech prophecies and advertisements, the surface glow of technological optimism expressed in Tsukuba, in Seville, in Knoxville, in New Orleans, and elsewhere only partly conceals the underlying anxiety about the future and the ultimate inability to restore the more understandable hopefulness of earlier world's fairs.

Henry Adams, the great American historian and man of letters, was profoundly moved by his visit to the Paris exposition of 1900. At a time when so many other Americans, among others, predicted ever greater technological and in turn social progress, Adams expressed grave doubts about the future based in part precisely upon his experience at that world's fair. As he wrote in *The Education of Henry Adams* (1907):

Until the Great Exposition of 1900 closed its doors in November, Adams haunted it, aching to absorb knowledge, and helpless to find it. He would have liked to know how much of it could be grasped by the best informed man in the world. . . . Historians undertake to arrange sequence – called stories, or histories – assuming in silence a relation of cause and effect. These assumptions, hidden in the depths of dusty libraries, have been astounding, but commonly unconscious and childlike; so much so, that if any captious critic were to drag them to light, historians would probably reply, with one voice, that they had never supposed themselves required to know what they were talking about. Adams, for one, had toiled in vain to find out what he meant. . . . Where he saw sequence, other men saw something quite different, and no one saw the same unit of measure. . . . Satisfied that the sequence of men led to nothing and that the sequence of their society could lead no further, while the mere sequence of time was artificial, and the sequence of thought was chaos, he turned at last to the sequence of force; and thus it happened, that after ten years' pursuit, he found himself lying in the Gallery of Machines at the Great Exposition of 1900, his historical neck broken by the sudden eruption of forces totally new.[68]

Would that any future world's fairs, building on the message of Vancouver's Expo '86, might have a similar profound impact on at least a few of their visitors.[69]

IV

Finally, high tech promotes "technological literacy" but in so doing displays a narrow, ahistorical conception of literacy and suggests a defensiveness about what genuine technological literacy reveals about technology's past, present, and future. The indifference to technology's past here is no worse – but no better – than the obsession with technology's past in high tech advertising. Like the high tech prophets, though, the promoters of technological literacy ultimately cannot escape the historical legacy of technology's mixed blessings.

More than three decades after C. P. Snow identified and lamented the growing gap between the "two cultures" – the science and the humanities – technological literacy is being widely hailed as the long-sought bridge between them.[70] Curricular reforms at several leading liberal arts colleges, supported by the private Alfred Sloan Foundation, have been introduced to institutionalize technological literacy. They include courses and research projects ranging from computer-generated art and music, to bridge, cathedral, and skyscraper design, to game theory and national security, to the social and psychological effects of noise pollution. And these so-called New Liberal Arts programs are merely the forerunners of anticipated nationwide educational reforms intended to restore America's economic as well as technological supremacy. Not since the upheavals in American education after the 1957 Sputnik satellite launching has so much concern been expressed about technical (and scientific) education.[71] In the representative words of two technological literacy advocates, "technological development" is accelerating so fast that "fewer and fewer people will understand these changes." This could create a "technologically illiterate nation." Shades of Toffler's *Future Shock*! Thus everyone must "understand technology if we are to function as citizens in such roles as voters, workers, employers, consumers, and ... family members."[72]

Yet the concept – and practice – of technological literacy, as evidenced by these initial efforts, may be more problematic than its advocates realize. Technological literacy is attended with several ironies that call into question its ultimate utility for its intended beneficiaries: those outside the ranks of engineers and other technical experts who wish to know more about and participate more fully in our avowedly technological society, particularly future decision-makers, not simply ordinary citizens.

The first irony of technological literacy is that, contrary to what one might expect, the term does not mean simply becoming acquainted with technology as hardware, that is, as computers, word processors, faxes, and the other tools of the contemporary revolution in electronics and information processing. Technological literacy is only partly a vocational enterprise, only partly a matter of learning the new alphabet of that revolution. But insofar as it is such, only the most diehard humanists could reject this reform. In fact, many self-proclaimed humanists have openly embraced the hardware.

In truth, however, as represented by the New Liberal Arts agenda, technological literacy also means becoming familiar with the nature of technology itself, or what I like to call "technology as software": its principles, functions, and values. Here the hardware is secondary, and technology is appropriately deemed as much an intellectual as a material phenomenon. Hence the emphasis within the New Liberal Arts on basic engineering principles, applied mathematics, computer models, and other aspects of "quantitative reasoning" – to use one of its favorite terms. Not surprisingly, the persons most often requiring such enlightenment are those in humanities, some of whom still pride themselves on their ignorance of – if not hostility toward – technology. That such persons have usually been the most fervent advocates of forcing engineering (and science) students to acquire a liberal arts education is the second irony of technological literacy; for what they demand of others they frequently refuse to do themselves.

Yet those humanists' traditional assumptions about technology being separate(d) from its human creators and potentially harmful to their interests, retain a certain validity today, as Americans' long-standing faith in technical experts and in their allegedly "fail-safe" systems declines (even in the face of the temporary resurgence of technological utopianism produced by the Persian Gulf

War). Certainly technology has always been and continues to be very much a human creation, as the New Liberal Arts programs properly try to demonstrate. However, the fundamental fear on the part of such humanists that contemporary forms of technology – whether robots or computers or nuclear weapons – will get out of control is not without basis. Indeed, the more nonexperts learn about modern technology the more they may become concerned as much as converted. Their much-vaunted new ability to "do" technology rather than, as in many older Science, Technology, and Society Programs, merely talk about it, may thus prove harmful to technology's own advance.[73] This is the third irony of technological literacy.

The fourth irony is that many scientists are also in need of enlightenment about the nature of technology. As Snow himself recognized, scientists can be as patronizing toward engineers and other technologists as any self-proclaimed humanists. He thus deemed technology a separate or third culture – a point widely ignored by commentators on his work.[74] Appropriately, the New Liberal Arts have illuminated the profound differences between science and technology and have suggested that technology is not merely the application of scientific principles but is rather, as noted, a highly intellectual enterprise in itself: one oriented toward design. As Princeton civil engineer David Billington, perhaps the leading practitioner of the New Liberal Arts, has argued, "Science is discovery, engineering is design. Scientists study the natural, engineers create the artificial. Scientists create general theories out of observed data; engineers make things, often using only very approximate theories." Engineers' "primary motive for design is the creation of an object that works."[75] That such differences exist between science and technology surely bolsters the case for including technology in the liberal arts curriculum as an independent intellectual construct, along with science. Not all scientists, though, would be comfortable with broadening the liberal arts in this way.

The fifth irony of technological literacy, however, is that some of its advocates appear to desire not just the discussion but the outright acceptance of technological thinking and values by humanists: above all, that pervasive assumption among engineers and technologists generally that the fundamental problems of the world are technical in nature and that those problems both have been and are being routinely solved by technology. True, as Billington observes, engineers often see more than one solution to a problem; and problem solving in itself can certainly be a creative endeavor. But the idea that there may *not* be a solution to every given problem, technical or not, is alien to engineers, just as it is to the high tech prophets, advertisers, world's fair/theme park designers, and their collective corporate employers and patrons. Writings that suggest this, such as Franz Kafka's *The Trial* and *The Castle*, are deeply disturbing to engineering students (and their mentors), as I have learned from my own teaching experience. Not surprisingly, one still finds genuine technological utopians among the advocates of robotics, genetic engineering, space colonies, nuclear power, and, of course, "Star Wars." Nor is it accidental that engineering students assigned Aldous Huxley's *Brave New World* often misinterpret the work as a genuine utopia, a technocrat's dreamworld with themselves as the Alphas, or leaders of society. The skeptical, often critical view of the world and of its imperfect inhabitants that one commonly associates with the humanities – from Greek tragedies to Shakespearean plays to antiutopian novels – is thereby threatened by this extension of technological literacy. By contrast, civil engineer and popular writer Samuel Florman, a prominent advocate of integrating engineering with the "old" liberal arts, repeatedly emphasizes the "tragic view of life" gleaned from a more traditional humanities curriculum and its value to engineers and nonengineers alike. Hardly an opponent of technological progress, Florman nevertheless tempers his optimism with an appreciation of mankind's limitations that his own exposure to the humanities has taught him. Florman thus would not, I believe, favor transforming the liberal arts curriculum in the new directions outlined here.[76]

Significantly, as reported in the educational press, certain of the initial Sloan-supported efforts have been criticized by the foundation for insufficient attention to quantitative reasoning and other "technological modes of thought" in their teaching and research.[77] The assumption that the most fundamental issues and values can be quantified in one way or another – and that those not quantifiable are not truly important – equally distinguishes technology from the traditional liberal arts, where such assumptions are invariably rejected as naive and shallow. Furthermore, the kind of precise, analytical thinking found in technology may not apply as readily to the liberal arts as some of its advocates suggest. Consequently, no matter how warmly humanists embrace technology, this profoundly different worldview remains.

The sixth irony is that technological literacy may not be as essential to the daily functioning of our high tech society as its advocates would have us believe. For all the hype about the acknowledged need for ordinary citizens, but especially nontechnically trained decision-makers, to understand better the world in which they live, it has yet to be demonstrated, as one curriculum development expert concedes, "exactly how much knowledge and skill in scientific and technical fields an individual needs to work as an accountant, a construction worker, a clerk, a secretary, or a teacher" – or, for that matter, as a corporate executive or a government official.[78] If, as in the exemplary case of computers, high tech has become so "user friendly" that one hardly requires extensive technical expertise to operate them successfully, what is the compelling case for technological literacy in the broader sense espoused by its advocates? Is it really true that, as a leading New Liberal Arts advocate assumes, "If people understand what's going on inside a computer or TV, they will enjoy it more"?[79] Could one not argue that high tech's very advances in cases like computers (and televisions) have undermined rather than advanced technological literacy's utility?

The seventh irony is that an appreciation of technology's own past might well make one cautious about technological literacy's future. Where, however, the humanities provide a sense of history, most advocates of technological literacy seem to have little interest in history, including the history of technology. (The teaching and scholarship of Billington on machines and structures, two very different sides of technology, is a notable exception.) For them, as for other high tech visionaries, technology is only future oriented, and what is past is literally "bunk."[80] In fact, technology has a rich history, as its scholars have made abundantly clear in recent decades, and technological literacy ought to be broadened to encompass that history. For that matter, the very idea of spreading technological literacy in America can be traced back at least as far as the nineteenth-century mechanics institutes that provided basic technical instruction to working-class citizens. Are the proponents of technological literacy aware of these antecedents? Or do they see themselves as the originators of this crusade?

Equally troublesome, the majority of those few technological literacy crusaders who do incorporate history in their teaching and research appear to believe that history shows technology to have always been a positive and progressive force, only solving problems, never creating them; this naturally leads to technological utopianism. Yet the history of technology, far from being an exemplar of the overly optimistic Whig theory of history, invariably reflects technology's mixed impact on and reception by virtually every society, from the most technologically primitive to the most advanced. Hence the real lesson to be learned from the serious study of technology's past is that technological progress and social progress do not necessarily go hand-in-hand but often conflict. The same lesson, of course, can be derived from the serious study of such contemporary technological advances as pesticides, organ transplants, nuclear weapons, and space shuttles. A course or program in technological literacy that suggested this might nevertheless risk loss of foundation, corporate, governmental, or even college and university support. So might a course or program that led newly enlightened citizens to control (their) technology by, say, opposing the nuclear power plants or nuclear weapons they now truly understood. Has the technological literacy crusade sufficient "self-confidence" – to use another of its favorite terms – to accept serious questioning of its worldview? Or to consider seriously, in the words announcing a recent national conference on technological literacy, the "ethical and value implications" of technology, much less the political implications?[81]

The eighth irony of technological literacy is that its advocates' democratic rhetoric does not quite mesh with the elitism inherent in both the major engineering schools and the prestigious liberal arts colleges that have nearly monopolized public and private funds for the New Liberal Arts. This situation does show signs of changing – as with the excellent Sloan-supported program at the State University of New York at Stony Brook, a large public and to this extent nonelitist institution – but the persistent emphasis upon (re)educating present and future decision-makers loses some of its appeal outside those charmed circles.[82] (Subsequent Sloan funding at twelve historically black colleges is another step in a more democratic direction.) Can technological literacy be as effectively applied to mass higher education as it has been to date to elite institutions? For that matter, can technological literacy "trickle down" to mass secondary and elementary education as readily as its advocates suggest it can? No less important, can the managerial ethos endemic to the training of future engineers be tempered in the direction of sharing responsibility for technology's future with ordinary citizens? And can characteristically indifferent engineering and other technical students be stirred politically or socially?

The ninth and final irony is that the very push for technological literacy ultimately reflects ambivalence, or at least anxiety, on the part of its proponents about technology's future. This is not to deny their superficially optimistic views and values, as elaborated upon above, but rather to suggest their underlying concern as to why such optimism is not widely shared among other educated persons. In varying ways the same concern accounts for the overcompensatory zeal of high tech prophets, advertisers, world's fair/theme park designers, and their corporate sponsors. Admittedly, technological literacy efforts *are* winning converts away from the hostility toward technology (and science) found in some of the Science, Technology, and Society (STS) programs of the late 1960s and early 1970s. These efforts might, in fact, sway certain programs in the opposite direction, toward what Langdon Winner has called HSTS – "Hooray for Science, Technology, and Society."[83] Nevertheless, if technological literacy is truly to bridge the three separate cultures Snow described without simultaneously threatening the humanities and alienating the sciences, it must somehow incorporate a critical, historical perspective. Just as being literate in a conventional sense today entails more than being able to read and write – it now also means being able to function in society and to develop one's knowledge and potential – so being technologically literate should properly entail more than being comfortable with the hardware or even the software.

V

With technological literacy, then, as with prophecies, advertising, and world's fairs/theme parks, contemporary high tech cannot let go of the past as it otherwise passionately embraces the present and the future. It is as if, deep down, high tech's promoters fear that the world of tomorrow may prove considerably less than utopian, their public stance notwithstanding. In this regard let me add a concluding point. The eminent philosopher of science Nicholas Rescher has examined the findings of leading pollsters and concluded that Americans and, by extension, citizens of comparably industrialized societies generally find technological advances insufficient in themselves to constitute perceived increases in personal happiness. By and large, people do not oppose advances that enhance domestic comforts or contribute to the national welfare. But they simultaneously yearn for the "good old days," when, despite these advances, life was supposedly happier; and they simultaneously desire ever more such advances to meet the ever rising expectations created by the most recent advances. Hence a terribly complicated, even contradictory, set of assumptions colors conventional Western attitudes toward technological progress and social progress.[84] What most concerns Rescher, however, is the growing tendency to blame technology and science rather than ourselves for not producing the expected increase in personal happiness. This concern is surely legitimate although, I suggest, exaggerated by him and others. The deeper problem is determining what to do next, once it is agreed that, in Rescher's words, "science and technology cannot deliver on the $64,000 question of human satisfaction ... because, in the final analysis, they simply do not furnish the stuff of which real happiness is made."[85] This is what led me to that prospect of a technological plateau noted at the outset, the prospect partially undermined by the resurgence of technological utopianism during and after the Persian Gulf War.

Contemporary high tech is thus replete with many ironies of the unanticipated consequences of technological progress along the route to technological utopia. That such consequences are painfully familiar to those truly knowledgeable about technology's past is yet another irony. So long as its various promotional enterprises turn a profit and generate positive public relations, high tech may not care what ordinary citizens, much less cultural critics, feel about these dilemmas. Yet ordinary citizens, not simply intellectuals, may care a good deal more than high tech and its promoters believe (or allow themselves to believe). The ideological and other ends being served by unjustified technological optimism may turn out to be every bit as questionable as those being served by technological pessimism in other quarters. However successful its individual technologies, then, high tech is lacking in the very historical consciousness that would in turn temper its optimism and thereby, most ironically of all, perhaps strengthen its appeal.

Notes

1. The two principal collections of primary historical sources on public attitudes toward American technology reflect a fundamentally hopeful outlook, an outlook only periodically tempered by negative developments akin to those listed above. See Carroll W. Pursell, Jr. (ed.), *Readings in Technology and American Life* (New York: Oxford University Press, 1969), and Thomas P. Hughes (ed.) *Changing Attitudes toward American Technology* (New York: Harper and Row, 1975). Hughes' introduction and conclusion are especially illuminating in this regard. By contrast, the now infamous nationally televised speech of President Jimmy Carter of July 15, 1979, about the nation's "malaise," reprinted in the *New York Times* of July 17, 1979, is notably simplistic in its assumption of an unchanging American faith in technological progress until quite recent times.
2. See Howard P. Segal, "Let's Abandon the Whig Theory of the History of Technology," *The Chronicle of Higher Education* 18 (July 19, 1979), 64.
3. See W. Warren Wagar, *The Next Three Futures: Paradigms of Things to Come* (Westport, Conn.: Greenwood Press, 1991), p. 26.
4. Alvin Toffler, *Future Shock* (1970; rpt. New York: Bantam Books, 1971), pp. 9, 11.
5. *Ibid.*, p. 2.
6. *Ibid.*, pp. 25, 266, 124, 125.
7. *Ibid.*, pp. 7, 1.
8. *Ibid.*, p. 5.
9. *Ibid.*, pp. 478, 470.
10. *Ibid.*, p. 391.
11. Alvin Toffler, *The Third Wave* (1980; rpt. New York: Bantam Books, 1981), p. 349. In 1975 Toffler published *The Eco-Spasm Report* (New York: Bantam Books), a short book that succinctly provided the same basic thesis as *The Third Wave* but called the phenomenon "Super-industrialism."
12. *Ibid.*, pp. 10, 182, 11.
13. *Ibid.*, pp. 424, 442.
14. *Ibid.*, pp. 391, 357.
15. Edward Cornish quoted in review of *The Third Wave* by Jerry Adler, *Newsweek* 95 (March 31, 1980), 86.
16. Toffler, *The Third Wave*, p. 1.
17. Alvin Toffler, *Previews and Premises: An Interview with the Author of "Future Shock" and "The Third Wave"* (New York: William Morrow, 1983), p. 89.
18. *Ibid.*, pp. 208, 213, 214, 215.
19. Alvin Toffler, *Powershift: Knowledge, Wealth, and Violence at the Edge of the 21st Century*

(New York: Bantam Books, 1990), pp. 3, 11, 75.

20. *Ibid.*, pp. 18, 19.

21. *Ibid.*, p. xix.

22. John Naisbitt, *Megatrends: Ten New Directions Transforming Our Lives* (New York: Warner Books, 1982), p. 8.

23. Naisbitt disagrees with Toffler regarding the appeal of "electronic cottages," claiming, with some justification, that most people want to go to offices and other workplaces and to associate with one another, not work at home. See Naisbitt, *Megatrends*, pp. 35–36.

24. Ben Bova, review of *Megatrends*, *Washington Post Book World* 12 (October 17, 1982), 1.

25. Naisbitt, *Megatrends*, p. 1.

26. Bova, review of *Megatrends*, p. 1.

27. Naisbitt, *Megatrends*, p. 252.

28. Naisbitt and Patricia Aburdene, *Megatrends 2000: Ten New Directions for the 1990's* (New York: William Morrow, 1990), p. 11. In 1985, Naisbitt and Aburdene published *Re-inventing the Corporation: Transforming Your Job and Your Company for the New Information Society* (New York: Warner Books). Where *Megatrends'* purpose was to describe the "new world" outlined above, this book's mission was to tell corporate executives and others "what to *do* about those and other new changes which compel us to rethink every aspect of corporate life." (p. 4). Perhaps *Megatrends 2000* will soon be followed by a similar corporate operations manual.

29. Naisbitt and Aburdene, *Megatrends 2000*, p. 311.

30. *Ibid.*, p. 309.

31. On Infomart, see Toni Mack, "Crow's Feast," *Forbes* 134 (December 17, 1984), 216, 220; Peter Petre, "Selling: Computer Marts: A New Way to Hawk High Tech," *Fortune* 111 (February 4, 1985), 64; Jim Hurlock, "Information Processing: Slow Starters – Or White Elephants? High-Tech Trade Marts Haven't Drawn Many Crowds – Or Exhibitors – But Backers Remain Optimistic," *Business Week* 2933 (February 17, 1986), 78; Lisa M. Keefe, "The Soft Sell and the Crystal Palace," *Forbes* 139 (June, 15, 1987), 130–131; Dennis Eskow, "Infomart: A Boon for Busy IS Managers," *PC Week* 6 (September 25, 1989), 81, 85; and Peter H. Lewis, "The Executive Computer: Connectivity Comes to Life in a Technology Supermarket," *Sunday New York Times*, September 29, 1991, Business, p. 11.

32. Quoted in 1985 or 1986 untitled Infomart brochure.

33. (no author) "The Crystal Palace Story," *Infomart Directory First Quarter 1992*, p. 2.

34. See Paul Goldberger, "Javits Center: Noble Ambition Largely Realized," *New York Times*, March 31, 1986, pp. B1, B2. On the Crystal Palace itself, see Folke T. Kihlstedt, "The Crystal Palace," *Scientific American* 251 (October 1984), 132–143.

35. Letter to author from Janice Collins, Account Executive, Tracy-Locke/BBDO Public Relations, September 17, 1986.

36. See (no author) "Videoconferencing: A New Mindset in Corporate Communications," *Infomart Magazine* (First Quarter 1992), 14–15.

37. On Leonardo da Vinci and advertising for Xerox and other high tech companies, see Brian Moran, "Leonardo Sighs: High-Tech Companies Adopt Artist's Image," *Advertising Age* 57 (October 6, 1986), 118, and (no author) "What's New Portfolio," *Adweek* 26 (January 16, 1989), 34. Moran's article discusses the competition among Xerox and two other high tech companies using Leonardo in their ad campaigns to claim originality in their historical appropriations, with the two others also claiming greater historical sensitivity than Xerox.

38. On the models of inventions from this traveling exhibit and their use in later museum exhibits, see (no author) "Exhibit Showcases da Vinci's Inventions," Associated Press article in *Bangor Daily News*, June 27–28, 1987, p. 12. On the builder of these models, Roberto Guatelli, see

John Culhane and T. H. Culhane, "The Master Modeler: Leonardo as a Lifelong Inspiration," *Science Digest* 93 (December 1985), 66–71, 86–87.

39. Richard Terrell quoted in Eskow, "Infomart: A Boon for Busy IS Managers."

40. On the Michelangelo computer virus, see Josh Hyatt, "Leading Edge Attacked By Rogue Virus," *Boston Globe*, January 29, 1992, pp. 65–66; *idem*, "Computer Killers," *Boston Globe*, March 3, 1992, pp. 35, 45–56; and *idem*, "Happy Birthday, Michelangelo," *Boston Globe*, March 7, 1992, p. 25.

41. See Lawrence M. Fisher, "Apple to Give the Public Its First Look at Newton," *New York Times*, May 29. 1992, p. D3. On Isaac Newton, apples, and gravity, see I. Bernard Cohen, *The Birth of a New Physics*, 2nd ed. (New York: Norton, 1985), pp. 41.

42. These ads appeared in, among other places, several issues of *Newsweek* in 1990 and 1991.

43. On Charlie Chaplin's Little Tramp and advertising, see (no author) "Critic's Corner: A Rose for IBM," *Advertising Age* 53 (August 30, 1982), M-29; Daniel Burstein, "Using Yesterday to Sell Tomorrow: How the Unlikely IBM–Charlie Chaplin Marriage Came to Be," *Advertising Age* 54 (April 11, 1983), M-4, M-5, M-48; Bob Marich, "Fortune Jabs Little Tramp," *Advertising Age* 55 (May 21, 1984), 1, 96; Patricia Winters, "Little Tramp Ends Sting As IBM Front Man," *Advertising Age* 58 (April 6, 1987), 80; and Jon Lafayette, "IBM's Little Tramp Returns to Ads," *Advertising Age* 62 (September 2, 1991), 5. Marich's article discusses the attempts by two small computer companies to use the Little Tramp in ad campaigns that followed IBM's. See also Charles J. Maland, *Chaplin and American Culture: The Evolution of a Star Image* (Princeton, NJ: Princeton University Press, 1989), pp. 362–370.

44. Coincidentally, the Computer Museum's new "People and Computers" exhibition has both a video and a photo of but, alas, no comment on the television version of the IBM Little Tramp advertisement. On the Computer Museum overall, see Howard P. Segal, "Exhibition Review: Computers and Museums: Problems and Opportunities of Display and Interpretation," *American Quarterly* 42 (December 1990), 637–656.

45. Among other places, the Sun Microsystem advertisement appeared in the *Boston Globe*, February 24, 1992, p. 9; the Storer Communications telegraph operator advertisement in *Advertising Age* 54 (May 9, 1983), 30–31; the Storer Pony Express rider advertisement in *Advertising Age* 54 (July 18, 1983), 54–55; and the Lockheed advertisement in a 1988 issue of *Newsweek*.

46. *El Al* 41 (January–February 1992), 2.

47. On Jacob Bigelow's *Elements of Technology*, see Howard P. Segal, *Technological Utopianism in American Culture* (Chicago: University of Chicago Press, 1985), pp. 78–81, 180 n. 7, 208 n. 15, 208 n. 16, 209 nn. 19, 20, 21, 210 n. 28.

48. Interestingly, R. L. Polk, a high tech company, had an advertisement using Leonardo and his proposed inventions with the caption "State of the Art Technology" in *Advertising Age* 58 (May 18, 1987), 8.

49. See Jeffrey L. Meikle, *Twentieth-Century Limited: Industrial Design in America, 1925–1939* (Philadelphia: Temple University Press, 1979), chap. 9. On the history of world's fairs, see the very useful John E. Findling (ed.), *Historical Dictionary of World's Fairs and Expositions, 1851–1988* (Westport, Conn.: Greenwood Press, 1990). Not surprisingly, perhaps, Findling's preface, p. xiii, implies a belief that world's fairs not only will persist into the twenty-first century but also that they should.

50. See Frank Vizard, "Electronics: The Flavor of CES," *Popular Mechanics* 167 (April 1990), 42, 44, and Jerry O'Neill, "Show Reports: Photokina," *Lighting Dimensions* 13 (January 1, 1989), 80, 86–87.

51. See Steven Greenhouse, "Playing Disney in the Parisian Fields," *Sunday New York Times*, February 17, 1991, Business, p. 1, 6; Sarah Catchpole, "Foreign Journal: Paris Chic Hits

Mickey Mouse Code," *Boston Globe*, February 13, 1992, p. 2; and Ronald Koven, "French Debate 'Better Mousetraps,'" *Boston Globe*, April 13, 1992, p. 5. The French intellectual credited with the Chernobyl reference varies with each story.

52. Michael L. Smith observes a similarity in both layout and purpose between EPCOT and the 1939–1940 New York World's Fair. See his "Back to the Future: EPCOT, Camelot, and the History of Technology," in Bruce Sinclair (ed.), *New Perspectives on Technology and American Culture* (Philadelphia: American Philosophical Society, 1986), p. 70.

53. See Priscilla Painton, "Fantasy's Reality," *Time* 137 (May 27, 1991), 52–59.

54. Mike Wallace, "Mickey Mouse History: Portraying the Past at DisneyWorld," *Radical History Review* 32 (March 1985), 49.

55. Marty Sklar quoted in Jennifer Allen, "Brave New EPCOT," *New York* 15 (December 20 1982), 41, 43. Mickey Mouse, Donald Duck, and other Disney cartoon characters *are* banned from EPCOT in the name of seriousness. On the genuinely serious crop research at EPCOT in the Kraft-sponsored Land pavilion, see John Rothchild, "EPCOT: It's a Stale World After All," *Rolling Stone* 403 (September 1, 1983), 38; David Hall, "The Green Earth – Care of Disney," *New Scientist* 103 (July 26, 1984), 41; and Judie D. Dziezak, "EPCOT Center Adds New Exhibit: A Plant Biotechnology Laboratory," *Food Technology* 42 (December 1988), 110, 114–115, 117. On EPCOT's semieducational exhibits and computer terminals, see Elting E. Morison, "What Went Wrong With Disney's World's Fair," *American Heritage* 35 (December 1983), 77.

56. Jeff Kramer, "Recreating Make-Believe," *Boston Globe*, June 14, 1992, p. 2. The architect quoted is Craig Hodgetts.

57. On these professional theme park and world's fair organizers, see Mark Zieman, "Jinxed Effort: New Orleans Prepares For Its World's Fair and Plenty of Criticism," *Wall Street Journal*, March 26, 1984, 1, 12, and Louise Zepp, "IAFE Sets Goal At 500 In Membership Drive," *Amusement Business: The International Newsweekly for Sports and Mass Entertainment* 102 (May 7, 1990), 1, 52. IAFE refers to the International Association of Fairs and Expositions.

58. Ray Dykes, "How PR Packed 'Em In At Expo '86: Interview with George Madden," *IABC Communication World* 4 (February 1987), 13.

59. Findling's Introduction does note that in the earliest world's fairs held in the United States, American participation, unlike that of European countries, "was undertaken by private means; there was much resistance to government support among Americans used to a sense of isolation and nonentanglement with the Old World" (Findling, *Historical Dictionary*, p. xviii.)

60. See (no author) "Debate Flares on U.S. Role at a World's Fair," *New York Times*, February 12, 1990, p. B12; Michael K. Frisby, "Budget Woes Imperial U.S. Pavilion for '92 Fair," *Boston Globe*, November 14, 1991, p. 22; Jonathan Kaufman, "Money Constraints Leave U.S. With Modest Showing at Expo '92," *Boston Globe*, April, 20, 1992, pp. 1, 9; and Alan Riding, "Seville Journal: The World's Fair Opens, but Where's Columbus?" *New York Times*, April 21, 1992, p. A4.

61. Quoted in 1986 *Infomart Fact Sheet*, no page number, prepared by Tracy-Locke/BBDO Public Relations.

62. See William E. Schmidt, "The Desolate Legacy of Knoxville's World's Fair," *New York Times*, May 18, 1984, p. A10; Dale Russakoff, "'82 World's Fair Pavilion: From Dazzler to Disaster," *Washington Post*, June 9, 1987, p. A3; Zieman, "Jinxed Effort," pp. 1, 12; and Wayne King, "Failed Fair Gives New Orleans a Painful Hangover," *New York Times*, November 12, 1984, p. A16.

63. See, as an example of this, (no author) *Selling the World of Tomorrow*, catalog for 1989–1990 exhibit of same name on the 1939–1940 New York's World Fair (New York: Museum of the City of New York, 1989).

64. David L. Groves, "Comparison of the 1982 and 1984 World's Fairs From the Participant Outcome Perspective," *World Leisure and Recreation* 31 (Summer 1989), 37.

65. Henry Raymont, 'A $6-million Shortfall for Columbus," *Boston Globe*, November 25, 1991, p. 11. On the Seville World's Fair, see also the articles cited in n. 60 above.

66. See John Barber, "Pavilions of Promise: The Best of Expo '86," *Maclean's* 99 (March 17, 1986), 24–26; Christopher S. Wren, "Vancouver Unwraps Its World's Fair," *Sunday New York Times*, April 20, 1986, Travel, p. 15; and Eleanor Wachtel, "Expo' 86 and the World's Fairs," in Robert Anderson and Wachtel (eds.), *The Expo Story* (Madeira Park, B.C.: Harbour, 1986), p. 29.

67. See Clyde Haberman, "Japanese See a 'Made in Japan' Future and Feel Reassured by That Vision," *Sunday New York Times*, May 5, 1985, p. A12.

68. Henry Adams, *The Education of Henry Adams: An Autobiography* (1907), partially rpt. in Hughes, *Changing Attitudes Toward American Technology*, pp. 168, 170–171.

69. A related plea is Langdon Winner, "A Postmodern World's Fair," *Technology Review* 94 (February/March 1991), 74.

70. See C. P. Snow, *The Two Cultures: And A Second Look* (New York: Cambridge University Press, 1964). The original version of the book, entitled *The Two Cultures and the Scientific Revolution*, appeared in 1960.

71. See, for example, Fred M. Hechinger, "About Education: Foundation Urges Drastic Change," *New York Times*, September 29, 1981, p. C4; James D. Koerner (ed.), *The New Liberal Arts: An Exchange of Views* (New York: Alfred P. Sloan Foundation, 1981); Edward B. Fiske, "Technology Gaining a Place in Liberal Arts Curriculums," *Sunday New York Times*, April 29, 1984, pp. 1, 20; John G. Truxal et al., "Liberal Learning and Technology: Building Bridges," *Change: The Magazine of Higher Learning* 18 (March/April 1986), 4–41; J. Ronald Spencer, *The NLA: Retrospect and Prospect* (New York: Alfred P. Sloan Foundation, 1988); John P. Brockway, *Technology and the Liberal Arts: Mixing Different Thought Patterns* (New York: Alfred P. Sloan Foundations, 1989); and Samuel Goldberg, "On Behalf of the Foundation," *NLA News* 8 (May 1992), 4–5.

72. Lee Smalley and Steve Brady, *Technology Literacy Test: A Report* (Menomonie: University of Wisconsin-Stout, 1984), p. 1.

73. On "doing" technology rather than merely talking about it in the context of the New Liberal Arts, see Morton Travel, "The Evolution of Vassar College's STS Program and Its Interaction With the 'New Liberal Arts,'" *Science, Technology, and Society: Curriculum Newsletter of the Lehigh University STS Program* 56 (November 1986), 4.

74. See Snow, *The Two Cultures*, pp. 29–33.

75. David P. Billington, "In Defense of Engineers," *The Wilson Quarterly* 10 (New Year's 1986), 89, 87. See also the comments in Fiske, "Technology Gaining a Place," of engineering professor Truxal of the State University of New York at Stony Brook regarding the differences between science and engineering, and John Nicholas Burnett (ed.), *Technology and Science: Important Distinctions for Liberal Arts Colleges* (Davidson, N.C.: Davidson College and Alfred P. Sloan Foundation, 1984).

76. See, for example, Samuel Florman, "Technology and the Tragic View, in his *Blaming Technology: The Irrational Search for Scapegoats* (New York: St. Martin's Press, 1981), pp. 181–193, and *idem*, "The Education of an Engineer," *American Scholar* 55 (Winter 1985–1986), 97–106.

77. See Judith Axler Turner, "Project to Include Technology in Liberal-Arts Curriculum Deemed Successful So Far, But Goal Remains Elusive," *The Chronicle of Higher Education* 31 (November 27, 1985), 20. The phrase "technological modes of thought" comes from Samuel Goldberg, "The Sloan Foundation's New Liberal Arts Program," *Change* 18

(March/April 1986), 14. See *ibid.*, pp. 14–15, and Nannerl O. Keohane, "Business as Usual or Brave New World? A College President's Perspective, *Change* 18 (March/April 1986), 28–29, for differing perspectives on the need to transform the entire undergraduate curriculum in order to accommodate Sloan's vision of appropriate technological literacy. Keohane's reference to Huxley's classic dystopia reveals a healthy skepticism of the New Liberal Arts missing in most other writings of administrators and faculty involved in the enterprise.

78. Dennis W. Cheek, *Thinking Constructively About Science, Technology, and Society Education* (Albany: State University of New York Press, 1992), p. 9. See also Fred M. Hechinger, "About Education," *New York Times*, August 3, 1988, p. B7, regarding the limits of scientific and, by extension, technological literacy.

79. Truxal quoted in Fiske, "Technology Gaining a Place," p. 1.

80. See, for example, the 1990 commencement address of University of Michigan President James J. Duderstadt, a fervent advocate of uncritical ahistorical technological literacy, before his institution's College of Engineering, abridged in *Michigan Today* 22 (June 1990), 14. See also the similarly uncritical, ahistorical approach to contemporary math and science education detailed in *Research News: University of Michigan Division of Research Development and Administration* 43 (Spring 1992). Ironically, as historian Bruce E. Seely has shown, the history of technology has for generations been deemed important within undergraduate engineering education at several progressive engineering schools and organizations. See his "The History of Technology and Engineering Education in Historical Perspective" (paper delivered at annual meeting, American Society for Engineering Education, Toledo, June 24, 1992).

81. The National Association for Science, Technology, and Society, which has held seven annual technological literacy conferences, and the now-defunct Council for the Understanding of Technology in Human Affairs, which published nine volumes of *The Weaver: Information and Perspectives on Technological Literacy*, have been much more sensitive to these nonquantitative concerns than the Sloan-funded programs and publications.

82. Truxal, director of this program, recognizes this dilemma. The teaching of technology, he says, "should not be restricted to students who will be future leaders; most of our students do not fall into this category" (Truxal, "Learning to Think Like An Engineer: Why, What, and How?" *Change* 18 (March/April 1986), 12).

83. See Langdon Winner, "On the Foundations of Science and Technology Studies," *Science, Technology, and Society: Curriculum Newsletter of the Lehigh University STS Program* 53 (April 1986), 6. See also *idem*, "Conflicting Interests in Science and Technology Studies: Some Personal Reflections," *Technology in Society* 11 (Fall 1989), 433–438; and Stephen H. Cutcliffe, "Science, Technology, and Society Studies as an Interdisciplinary Academic Field," *Technology in Society* 11 (Fall 1980), 419–425.

84. Nicholas Rescher, "Technological Progress and Human Happiness," in his *Unpopular Essays on Technological Progress* (Pittsburgh: University of Pittsburgh Press, 1980), pp. 3–22.

85. *Ibid.*, p. 19.